Structural Geomorphology

Geographies for Advanced Study

Edited by Professor Stanley H. Beaver, M.A., F.R.G.S.

Structural Geomorphology

J. Tricart

Director, Centre de Géographie Appliquée,
University of Strasbourg

Translated by

S. H. Beaver

Professor of Geography, University of Keele

and

E. Derbyshire

Senior Lecturer, Department of Geography,
University of Keele

Longman

Longman
1724-1974

LONGMAN GROUP LIMITED
London
and LONGMAN INC.,
New York
*Associated companies, branches and representatives
throughout the world*

English translation © Longman Group Limited 1974

Géomorphologie Structurale first published by the
Société d'Edition d'Enseignement Supérieur, Paris 1968
English edition first published 1974

ISBN 0 582 48462 6
Library of Congress Catalog Card Number: 73–86130

*Printed in Great Britain by William Clowes & Sons, Limited
London, Beccles and Colchester*

Contents

Contents

Contents

List of illustrations

List of illustrations

The geological maps of France referred to so frequently in the text are in the 1 : 80 000 series known as the *Carte géologique détaillée de la France*. These maps have been issued at various dates since 1870 and the series now totals some 258 sheets. The topographical base, printed in sepia, is the *Carte de l'Etat-Major* on the same scale.

Translators' preface

Structural Geomorphology is the first volume of a major work, *Précis de Geomorphologie*, the second volume of which, *Elements of Pedology and Hydrology*, appeared in French in 1972. The *Précis* is designed to provide both an introduction and a detailed systematic text of particular value to students reading for their first degree in geography. It is also intended to provide a comprehensive over-view of the science of geomorphology for specialists in cognate disciplines, notably geology, pedology, ecology and planning. Such is the organisation of the material that *Structural Geomorphology* may be used at a variety of levels depending on the degree of generalisation appropriate to the needs of the user. English-speaking readers coming to this work for the first time will find several distinctive and valuable features, amongst which might be mentioned the comprehensive reading guides which gather together a great deal of material rarely encountered in English-language texts of geomorphology; the impressive diversity of examples of structural landforms, many of them illustrated from Professor Tricart's own field-work; and a lucid introductory statement on the place of geomorphology within the Earth Sciences which contains a typically forthright assessment of the cycle of erosion according to W. M. Davis.

Professor Tricart's numerous other geomorphological texts, two of which have appeared in translation in the 'Geographies for advanced study' series, are listed in an appendix.

University of Keele
July 1973

S. H. Beaver
Edward Derbyshire

Introduction
The place of geomorphology among the earth sciences

Geomorphology, as the Greek roots of the word indicate, is the science of landforms. Its object is the study of the appearance and behaviour of the lithosphere, the earth's solid crust.

The nature of geomorphology

The surface of the earth's crust brings the solid mass of the globe into contact with the liquid hydrosphere and the gaseous atmosphere that surround it. Since man is by his very nature a terrestrial creature, it is the subaerial surface that is best known and most easily studied, but submarine relief is no less a part of geomorphology, and rapid advances are being made in this field.

The earth's crust, like any other surface affected by contact between different bodies, is moulded by the interaction of forces which originate on either side of it; this general physical law is basic to geomorphology. On the one side the crust is modified by internal forces, manifesting themselves in tectonic movements and vulcanicity; these forces control the differentiation between the continents and the ocean basins. On the other side are the external forces originating in the atmosphere and hydrosphere; these also play their part and interact with the internal forces. Atmospheric weathering affects rocks at their outcrop, and water, ice and wind transport the debris, here sculpturing patterns of erosion and there forming accumulations, not in the open air but also, especially, beneath the seas and oceans.

It is the interaction between the internal and external forces that controls the moulding of the earth's surface; and it is this which gives it its unique character. The surface of the moon, for example, now reasonably well known, has been fashioned solely by internal forces, in the absence of any atmosphere; this is the explanation of its resemblance to our earthly volcanic landforms, which themselves are largely manifestations of the internal forces.

The internal and external forces vary both in time and in space. Vulcanicity, for example, affects only certain parts of the globe at the present time, especially the borders of the Pacific Ocean, whilst other areas, such as

Scandinavia and the lands surrounding the North and Baltic Seas, are completely exempt. But this is only a synoptic glimpse in the long history of the earth; volcanic landscapes resulting from eruptions that have now ceased are to be found, for example, in the Eifel region of Germany. Similarly with tectonic movements. Intermittent earthquakes show that certain parts of the earth's crust are in a state of disequilibrium and subject to particularly intensive movement. Their distribution at the present time, however, is not the same as might have been observed 20 million years ago. The Alps, the Rhine graben, and the Limagne, which today are only affected by rare and very slight earthquakes, were at that time the scene of the violent and repeated shocks of a tectonic paroxysm.

The external forces fluctuate also. Coastlines have changed unceasingly during geological time. Encroachments of the sea—usually called transgressions—have subjected to submarine erosion surfaces that were formerly sculptured by subaerial agencies. Some 20 000 years ago the North Sea was almost all dry land, and peat bogs formed on the Dogger Bank before the area was submerged by the 'Flandrian transgression' which was caused by the rise of sea level resulting from the melting of the Würm or Weichselian ice sheets. Much of the relief of the continental shelves is of subaerial origin, resculptured to a greater or less extent by the recent advance of the sea. Only 15 000 years ago ice caps similar to that which now covers Greenland, enveloped Scandinavia and the Baltic, and the whole of northern and northeastern Canada as far south as the Great Lakes. The conditions under which the land surface of those regions evolved at that time were thus very different from those of the present day.

Both in time and in space, therefore, as well as in their nature and intensity, the manifestations of the internal and external forces are variable, as also is the degree of their interaction. Sometimes the internal forces are strongest, as when mountains are raised and depressions are invaded by the sea. At other times the external forces predominate, gradually reducing the mountains and filling the lower parts of the crust with sediment. There is thus a perpetual struggle, an uninterrupted morphological evolution, always in progress but never finished. If this were not so, the earth's surface would be dead, like that of the moon or an extinct star.

The object of geomorphology is thus the study of this surface of contact that is the earth's crust: the description and classification of the many forms and shapes that result from the ever-changing interaction of the opposed forces; the understanding of the mechanism of this interaction in order to comprehend the landforms that are generated by it, if necessary by penetrating backwards into time so as to trace the evolution of these forms and to place the present landscapes in their true historical context. Like the biological sciences, geomorphology is based on notions of evolution, of the interference of the various factors and their mutual adaptation. As with species, landforms are born, develop and then decline and disappear; some of them are relics or survivals, and the biogeographical term *endemism* springs to mind with regard to some of them. There exists a veritable ecology of landforms, that is an assemblage of circumstances that are necessary for the growth and development of this or that type of relief. But whereas in biology these circumstances are necessary but not sufficient, in geomorphology they are both necessary and sufficient, for although the landforms evolve they are not living things.

These analogies between biology and geomorphology are not just artificial. They are one aspect of the fundamental processes of nature, and we must have regard for them if we wish to comprehend these profound truths.

Our methods of reasoning share the same complexities as those of biology; and dealing as they do with antagonistic and ever-changing forces, they can only be dialectical. The Davisian theories, springing from geometrical reasoning nourished by imagination, have in fact checked and retarded the development of modern geomorphology. Instead of opening our eyes to the complex realities, they have lulled us into the security of a deceptive simplicity.

Before embarking on the study of the different types of landforms, it is necessary to take a broad general view of the interaction of opposing forces and to formulate a general model of causality in geomorphology that will guide our reasoning and place our observations in perspective.

General notions of causality in geomorphology

Figure o.1 shows in graphical form the causal relationships that govern geomorphology. This diagram has two conceptual bases, first that of evolution (time) shown vertically and ending in the present day, and secondly that of dynamism, of which the various aspects appear in the vertical columns, since their role is temporal. We have thus a system of interactions; these vary with time and their elements change in intensity from region to region, in conformity with the ideas expressed above. The vertical columns show the factors which are modified with the passage of time. Vertical arrows show how these factors bear on presentday geomorphology; horizontal arrows take account of the temporary or permanent interactions between these same factors. Lastly, the oblique arrows show the interactions that are produced over a period of time. In all cases the strength of the arrows is a rough indication of the intensity of the actions and interactions that they represent.

Although this model has been reduced to its utmost simplicity, and has been drawn with as much clarity as possible, it is by no means simple. Its complexity can but reflect the complexity of the causal mechanisms in geomorphology, and for this reason some further comments are necessary. We shall deal with each of the vertical columns in turn.

3

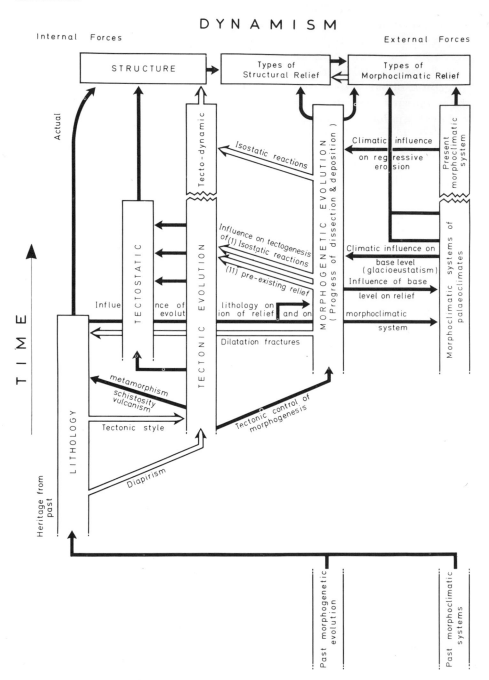

FIG. 0.1. Diagram of causal relationships in geomorphology

Internal forces

Three columns end at the level of the present day in the rectangle inscribed 'structure'. This shows that the geological structures that we see now, like those which might have been seen at any moment in the past, are the result of the combined action of several factors that depend more or less on the internal forces.

LITHOLOGY

Lithology, the nature of the rocks, is a direct product of the internal forces in the case of magmatic rocks like granite and basalt, that have been formed by plutonic and volcanic action. But a closer look at volcanic rocks shows that the situation is far from simple, for the form of the lava flows may be much modified by the nature of the land over which they pass; there is interaction between the internal forces that are responsible for the emission of the lavas, and other factors that control the nature of the underlying surface, which may be of sedimentary rocks almost unaffected by the internal forces. In the case of metamorphic rocks the interaction of internal and external geodynamic forces is even more obvious, for such rocks result from the modification of sedimentary rocks that are in contact with elements derived from the magma.

Lithology thus reflects the interactions of the internal and external forces. The latter are represented by the mineral components of rocks of magmatic origin and by the modifications suffered at depth by sedimentary rocks— which in turn are related to the tectonic movements that caused them to reach such depths. This is why, on the diagram, an oblique arrow leads from the 'tectonic evolution' column into 'lithology'. Along the course of this arrow we are reminded of the three ways in which tectonics influence lithology:

Metamorphism. This takes place when sedimentary rocks are depressed to a great depth; tectonic movements facilitate their penetration by material of magmatic origin, generally through the action of tensions which open up fissures into which the magma can more easily enter. Thus metamorphism often penetrates upwards to a greater extent in tensional structures such as domes and geanticlinees.

Vulcanism. This results from the localised rise of magma through fissures; the rise is so rapid that the magma has no time to get polluted by contact with the rocks traversed. All volcanic phenomena are associated with fracture zones, either the great faults that break up folded mountains into separate blocks or the faults that rupture the continental shields and platforms. Japan provides an example of the first case, East Africa of the second.

Schistosity. This results from the squeezing and stretching of sedimentary rocks under very great pressure. It is particularly clearly displayed in clays that have been subjected to foliation, but it is seen also in other rocks, especially marls and limestones. It only develops beneath at least 4 000 or 5 000 m of superincumbent strata, and occurs in geosynclinal structures. Other less extreme features are also discernible in some rocks. Diagenesis, which transforms mobile sediments into coherent rocks, results in part from the pressure exerted on the sediments by the deposits of younger age that accumulate on top of them, and this accumulation in turn results from slow subsidence. Friction along the tectonic discontinuities known as faults and thrust planes shatters the rocks and gives rise to fault breccias, sometimes called crush or tectonic breccias.

External influences on lithology are expressed through the medium of sedimentation. The sediments that are laid down at any given moment of time in any region are controlled by the following factors:

The nature of the rocks that are being eroded. Granite, for example, yields an abundance of quartz grains, whereas limestone yields almost none; limestone however yields much more calcium than does granite.

The processes by which the solid rocks are converted into rock debris that can be transported far from its point of origin. These processes vary greatly from one part of the world to another, especially in response to climate but also to a lesser extent in response to the nature of the relief. Where there is a dense vegetation cover the processes are mainly biochemical, and the products are chemical compounds in solution, and clays resulting from complex alteration mechanisms. On the other hand, where there is little or no vegetation, as in hot or cold deserts, and beneath ice, the rocks are rendered mobile through mechanical fragmentation and plucking. The products in this case are detrital, clastic materials, composed of unaltered rock fragments ranging in size from silt to boulders.

The mechanisms of transport. The materials weathered from the rocks at the surface are carried into sedimentation basins by running water, by ice or by the wind. They undergo, both in water and in the air, a sorting process that is a function of the strength of the current, and it is the finer material that stands the greatest chance of eventually reaching the sediment basin. While in the course of transportation by water, the materials are also subject to chemical action that alters them and extracts some substances in solution. Within glaciers and ice sheets there is no such sorting, but only abrasion that reduces part of the debris to rock flour.

Of these factors controlling the nature of the raw materials subsequently deposited to form sedimentary rocks, only one results directly from structure, and that is lithology. But the other two, the mechanisms of mobilisation and transport, are also integral parts of the morphogenetic system, controlled by climate and the nature of the relief. For this reason, in the lower righthand corner of the diagram, two columns represent morphogenetic evolution and the morphoclimatic system of a given period in the past; from the top of these columns runs an arrow leading to the base of the 'lithology' column. This is to indicate that the character of one sort of rock—the sedimentary—depends directly on the external forces that were operative at the time of the sedimentation. Subsequently, the internal forces play their part, in diagenesis, in crushing, in foliation—in a word, in the metamorphosis by which the rocks are variously modified. To this must be added vulcanism, bringing in other elements of internal origin.

TECTONICS

Tectonics are divided in the diagram into tectostatic and tectodynamic. In the usually accepted sense of the term tectonics, two ideas are confused: the disposition of the strata and the deformations that caused such dispositions. Thus, in referring to a folded mountain belt such as the Alps, we speak of the tectonics of thrusts to indicate the disposition of overthrust beds already in position, but we also use tectonics in referring to the complex movements that created such dispositions. In the first case we are concerned with a static disposition that has happened in the past and suffers no further modification: *tectostasy*. In the second case, on the contrary, we are concerned with the actual folding process: *tectodynamism*.

The two terms tectostatic and tectodynamic must be used, of course, in relation to time. They are only valid for a given moment of time in the history of a given region of the earth. Tectostasy refers to the actual disposition of existing strata (tabular, faulted or folded) and tectodynamism to the deformations that the rocks underwent at the given time period. Tectodynamism may be negligible—there may be an interval of tectonic calm. On the other hand, it may be very great, in which case we speak of a tectonic upheaval or paroxysm.

Tectodynamism often changes its character, so that, in the Jura for example, a phase of folding may follow a phase of faulting, or, as in the southern fore-Alps, several successive phases of folding may be characterised by folds of different orientation and aspect. Geologists call this superimposed tectonics. Each new paroxysm impresses its own brand of tectodynamism upon the pre-existing tectostatic circumstances. Another example of widespread occurrence may be drawn from those old massifs that were folded in the Palaeozoic; the folds are still visible because the metamorphism was not intense, and the tectostatic character is of a folded chain; subsequently the massifs were affected mainly by folds of less intensity and larger amplitude, accompanied now and again by fractures. After relatively recent uplift by tectodynamic forces, differential erosion acts on the tectostatic folds and in favourable lithological circumstances sculptures them into an appalachian type of relief. The distinction between

7

the static and dynamic aspects of tectonics at any given moment of time in the earth's history, but principally at the present time, thus plays an indispensable part in the development of ideas in structural geomorphology.

The successive tectodynamic episodes of the past are spoken of as the tectonic evolution which culminates in the tectodynamics of the present. This is why these two expressions form part of the same column in the diagram. Naturally, the state of affairs at any moment in the past constitutes a certain set of tectostatic circumstances, represented by certain dispositions of the rocks; and the durability of such dispositions varies with the rhythm and amplitude of the tectodynamic forces. In a period of tectonic calm the dispositions will be more long-lasting than during a paroxysm, when they will be in a state of continual disturbance. At any moment in the earth's history, the existing structure integrates the tectodynamic effects which have taken place since the emplacement of the affected strata. For this reason, in the diagram, a series of arrows leads from 'tectonic evolution' to 'tectostasy', and one of the arrows leads into the base of the latter column. In reality, we are dealing with effects that are more or less continuous, but a series of arrows is the most convenient form of diagrammatic representation.

Tectostasy and tectodynamism lead directly into present structure; but the arrows that indicate this are of different dimensions, for in fact in most cases the influences of tectostasy are much more important in morphogenesis than the influence of actual tectodynamism—though of course, the part played by each will vary from region to region. In a region that is going through a period of tectonic calm, tectostatics are much more important. Such is the case, for example, in the Paris Basin and southern England. On the contrary, in a region that is suffering a tectonic upheaval, like Japan, New Zealand or California, tectodynamism plays an important part that varies with the intensity of the paroxysm and is combined with the effects of differential erosion on the uplifted sections. Active faults, for example, give fault scarps that change with each earthquake. For this reason an arrow is shown leading from the 'tectonic evolution' column towards 'morphogenetic evolution'; it indicates the control of morphogenesis by tectodynamism. There are several interactions (indicated by arrows) between tectonic evolution and lithology. Lithology not only influences tectogenesis directly through the medium of diapirism but also acts in a less direct fashion.

Diapirism is a particular form of tectodynamism brought about by masses of salt (rock salt, potash etc.). Such salt masses, which are more plastic and have a density appreciably less than that of other rocks, have a tendency, when compressed, to seek an outlet to the surface by breaking through the rocks which cover them. They thus form domes, or diapirs, which often take on a mushroom form. The movements of the salt domes are relatively independent of other deformations and may occur even in times of tectonic calm. They are simply a product of lithology.

The mechanical properties of rocks also influence tectonics, for they control the degree of resistance that the rocks offer to tectodynamism. A sheet of paper does not fold in the same way as a piece of cardboard. Rock strata either fracture or bend, depending on their plasticity and on the intensity of the deforming stresses. For example, granite does not fold; it fractures. Clay strata containing salt, however, are not only very plastic but may also be slippery. Quite often they become stretched out, enabling the more rigid superincumbent strata to slide over them. In this way the rock layers become loosened, and differential movement culminates in the formation of overthrusts. On the other hand, as we have seen, tectonic history modifies the characteristics of rocks and so influences lithology. Thus certain tectonic features, such as open faults, permit volcanic emissions; and metamorphism results not only from the deep burial of sedimentary rocks but also from a whole series of conditions favouring the diffusion of magmatic elements through these rocks. The compression that produces schistosity is the direct result of the mechanical stresses to which the rocks are subjected in the course of tectonic deformation.

Important interactions thus occur between the various internal forces, while others are produced between the internal and the external forces. Both types have their part to play in the development of geological structures, but the internal forces are dominant. Indeed, the external forces operate only through the medium of the morphogenetic systems as they influence the formation of sediments. Though not negligible, this influence dates from the remote past, when palaeogeographical circumstances were very different from those of the present day. Nevertheless, structure is not entirely the product of internal forces, especially in regions of sedimentary rocks, where interactions between internal and external forces are apparent. But the predominance of the internal forces justifies the position of the terms 'internal forces' and 'structure' on the diagram.

External forces

Two major aspects of the working of the external forces may be distinguished, general morphogenetic evolution and morphoclimatic systems.

GENERAL MORPHOGENETIC EVOLUTION

General morphogenetic evolution responds to cosmic forces which are of two distinct kinds. In the first place there is gravity, which acts perpendicularly to the earth's surface and controls the downward movement of objects. In geomorphology, gravity sometimes acts directly, as in the formation of boulder screes, but more often indirectly through the agency of running water, glaciers and solifluction. It tends to lessen relief by removing debris from high places (slope summits, interfluves, mountain ranges) and depositing it in lower places (slope-foot colluvium, terminal moraines, *cônes de déjections*, deltas and submarine depressions where turbidity currents lose their loads). This reduction of relief by running water was the basis of W. M. Davis's theory of the cycle of erosion and peneplanation.

Secondly, there are the tangential movements that take place in the atmosphere and the hydrosphere. The winds and the currents in the sea and in lakes have an important horizontal component which is of some significance in morphogenesis. The winds, blowing over the surface of the earth, whip up dust and sand grains—a functional effect that is not fundamentally different from that exercised by a river on its bed, for both obey the same laws of fluid mechanics and only differ in respect of the value of the parameters representing the density of air and of water. But there is an important geomorphological difference, for running water obeys the law of gravity and always runs downhill, whereas moving air only responds to gravity in respect of its convectional movements and not in its turbulence when in contact with the earth's surface. Thus wind can overcome gravity and lift the debris it transports, as in the formation of sand dunes. The same effect happens on the sea shore; waves and tides may throw up beach material into strand lines which represent the level reached by the highest tides. Tidal currents and longshore drift cause the displacement of beach material laterally along coasts, an action more important than that of gravity.

Gravitational and tangential movements obey Newton's laws, which are among the basic principles of physics. They have been going on over the whole of the earth's surface ever since that surface came into being, and are thus the motive force behind the evolution of relief. They come into play wherever movable particles on the earth's surface are free to be transported. The wind, for example, with a power of movement much less than that of ice or water, can only act if the movable particles are not protected by vegetation and if they are small enough to be transported; the same applies to the movement of particles by a stream. This notion of competence can also be applied (excluding the protective vegetational cover) to the action of waves and longshore drift.

Gravitational and tangential movements may thus take place anywhere; they are of potential importance over the whole globe but are not necessarily effective everywhere. Gravitational movements are geomorphologically much more important than tangential ones, for they may affect all sorts of particles, from the chemical element dissolved in water running in a stream or percolating downwards through the rocks, to the boulder that rolls down the mountainside when its support is removed. The effects of gravitational movements are much greater than those of tangential movements, for the functional force of the wind is relatively slight, thus the former play the major role in the evolution of relief, except in limited spheres such as sand dunes and beaches. This is why there is a general tendency for relief to diminish with the passage of time. During the course of this evolution the surface rocks that yield the most mobile particles will be most rapidly eroded. This differential denudation is a function of lithology; through its medium the dissection of the surface becomes adapted to lithology and tectonic structure. Thus we get the sculpture of structural relief forms like cuestas, faultline scarps and the diverse forms derived from folded strata. But gravitational forces depend directly on the existence of inequalities in the earth's surface, and these result from tectonic action. A change of level of tectonic origin creates a certain morphogenetic potential, just as a waterfall offers hydroelectric potential. Under the influence of gravitational forces the change of level tends to be reduced, but before it can be reduced it must first appear. It is thus clear that interaction between tectonic movements and gravitational forces controls the evolution of tectonic relief.

For this reason the rectangle labelled 'morphogenetic evolution' in the diagram leads directly, by a broad arrow, into 'types of structural relief', into which there also comes an arrow from the compartment labelled 'structure'. Types of structural relief are the result of the interaction of gravitational and tectonic forces, working now in their tectodynamic aspect, now through tectostasy, within a certain lithological framework. This general morphogenetic evolution controlled by gravity interferes with tectogenesis and even with lithology. But these repercussions are apparently slight, and in the diagram they are shown by thinner arrows.

As for influences on tectodynamism, they are of two kinds:

1. Isostatic reactions are provoked on the one hand by the unloading that results from the erosion in an area subjected to vigorous dissection, and on the other hand by the arrival of large masses of debris in an area of accumulation such as a piedmont. Bearing in mind the high viscosity of the earth's crust, it is clear that such reactions will only take place when the forces pass a certain threshold. It has been calculated that reaction is only possible in the case of sharply delimited areas that are several dozen kilometres in size. For areas with more blurred margins the threshold is higher. The reaction in the case of an area subject to intense and rapid erosion, and so lightened, will be a tendency to uplift. But such erosion necessarily implies a previous tectonic uplift, so that isostatic reaction simply sustains an uplift of tectonic origin. The same thing happens, though in the opposite sense, in the case of depressed areas, which are subject to considerable accumulation.

2. The existence of pre-existing relief influences tectogenesis. For example, studies begun in the Jura and later extended to many other folded areas have brought to light modifications of tectonic deformations due to the resistance of the pre-existing folds. It is easy to see how an escarpment, subjected to tangential pressure, could advance laterally without encountering much resistance, whereas the rocks beneath the surface would be blocked against each other. When a shearplane occurs at the base of an escarpment—a condition easily produced in soft rocks like clays and marls—the whole feature may move forward to cover the rocks beneath

which are more resistant to the tangential pressures. In Provence, for example, many rock splinters are made up of the former crests of overfolds which have been thrust over the hollow which formerly lay at their foot. (For further details see Chapter 2.)

As for the repercussion of morphogenetic evolution on lithology, this is also a consequence of the removal of debris from regions undergoing erosion. It has been noted in many types of region, in high mountains and in narrow gorges, as in the Colorado, that cracks parallel to the walls develop as a result of the existence of the deep chasms. Some authors have ascribed the joints in granite to the progressive erosion of the overlying sediments. In very rigid rocks, the decompression resulting from erosion certainly tends to create fissures (called 'dilatation fractures') that are generally parallel to the surface; these in their turn facilitate further erosion.

As we have seen, gravity seldom works directly. Tangential forces, such as the wind, are hindered by a screen of vegetation. The intensity and the processes of morphogenetic evolution are influenced by a whole series of factors which are a function of the state of the earth's surface (vegetation, soil, superficial deposits etc.). Since climate controls these, we group them under the heading of 'morphoclimatic systems'.

MORPHOCLIMATIC SYSTEMS

The direct and indirect influences of climate play a fundamental role in morphogenesis. The vegetation cover protects the soil from certain mechanical actions such as the impact of raindrops, which are of great importance in the formation of runnels. In its turn, the decaying vegetation modifies the nature of the infiltrating solutions, and so not only influences soil formation but also chemical weathering such as karstification. The vegetation cover thus plays a large part in the process of transforming coherent rocks into debris that may subsequently be transported. Climate exerts its influence indirectly through the vegetation: each major vegetation type, forest, savanna, steppe and desert, is found in climatically determined areas. But there are also direct influences, through rainfall intensity, the strength and regularity of wind and frost. Polar coasts, for example, are protected from wave action during the whole or the greater part of the year because of the ice pack. For this reason we have begun to talk about climatic geomorphology: the study of the direct and indirect influences of climate in morphogenesis.

Corresponding to each climate there tends to develop a vegetation cover in equilibrium with it; this is termed the climatic climax. Under similar conditions morphogenesis is effected through the medium of a certain combination of processes, of which some are conditioned by climate and vegetation. In arid regions, for example, the sporadic rains attack the surface soil directly, for there is but a sparse protective vegetation cover. The development of runnels is easy, and this is everywhere the mode of surface sculpture. But chemical weathering, in the virtual absence of water, is at a minimum. Running water, moreover, can only erode friable rocks; when it comes up against the slightest solid obstacle it goes round it, leaving an upstanding edge or cliff. The erosion pattern is thus finely incised by innumerable small gulleys on the slopes, as is clearly seen in parts of the Sahara. Concentrated runnel development is not peculiar to deserts, far from it; but in the absence of a vegetation cover it develops without hindrance and gives the relief an unmistakable imprint. Indeed, it is the cornerstone of what we may call a 'morphoclimatic system'.

Morphoclimatic systems are specially important in conditioning the fragmentation of rocks, the transformation of coherent rocks or large boulders into fragments sufficiently small to be moved by the normal agencies of transport. In this way they control the whole process of morphogenetic evolution. Moreover, they exercise a strong influence on the way in which the gravitational forces operate. In the desert, for example, they work through the medium of sudden and concentrated torrents. In the savannas, the obstacles provided by plants hinder concentrated torrent formation, and the stream pattern remains diffuse for a long time. In periglacial regions, during the ice melt, with the soil still frozen to some depth, surface water that cannot infiltrate becomes liquid mud and gravitates slowly under its own weight; this is *solifluction*. Beneath the wet tropical forest, intense weathering and the loss of chemical elements in solution lead to subsidence and thus to another kind of soil flow. In each morphoclimatic system, the modes of operation of the gravitational and tangential forces are modified in a particular way. In the forest the wind merely waves the trees about, thus producing wind-fallen dead branches and trunks; but in the desert, provided there is sand, it heaps up dunes. The nature of the forces, by reason of their cosmic origin, is always the same; but the way in which they operate varies from one climate to another.

Morphoclimatic systems thus control the processes of landscape evolution. The vegetation cover and the climatic factors influence the stream regime and the ability of the rivers to transport debris, upon which depends their incision into uplifted regions. Vegetation and climate also control the fragmentation and alteration of rocks and the mobilisation of the debris on slopes, and so the way in which debris reaches the streams. If the debris is not too abundant or too large, it can be carried away, the rivers can incise their channels, and dissection makes progress. In the contrary case dissection is hindered. Thus there is continual interaction between morphoclimatic systems and morphogenesis.

An increase in slope, of tectonic origin, may set in motion a phase of erosion that involves the incision of the streams and then headward recession. This is a common phenomenon of morphogenesis. The incision of the stream beds increases the angle of the valley slopes, and so may modify the speed and even the nature of the sculptural processes. For example, earth slides may replace soil creep. Within the framework of the same climate, the same morphoclimatic system, morphogenetic evolution may thus be subject to modifications, notably the replacement of one process by another. There is thus a strong influence exercised by the base level of erosion on the morphoclimatic system. This is indicated in the diagram by a broad arrow.

Conversely, if the debris derived from the slopes is too abundant or too large, it may hinder the erosion of the riverbed and cause flooding. This is notably the case in some desert climates, under which fragmentation is intense but runoff deficient (e.g. the Peruvian coast at certain stages of the Quaternary). Or again, in hot, wet tropical areas, where the rivers carry but little abrasive load, outcrops of hard rock are difficult to erode and the rapids they create hinder headward erosion, so that the general evolution of the landscape is modified. This effect may be just as important as the preceding one, and is shown in the diagram by a similar arrow labelled 'climatic influence on regressive erosion'.

Of course, all the many ways in which the earth's surface is sculptured under various climates are directly and strongly influenced by the nature of the rocks. This is just as important as climate in the processes of fragmentation and alteration. For this reason a broad arrow leads from the 'lithology' column direct to 'morphoclimatic systems'.

Finally, in certain cases climate sets in motion isostatic reactions—for example, continental ice caps. During the glaciation, the weight of the ice depresses the affected area; but after the ice melt, relieved of its burden, the land rises again. Climate in this case has tectodynamic repercussions, particularly on the base level and general geomorphological evolution. Glaciations have also produced oscillations of sea level, regressions during the Ice-Ages and transgressions during the warmer periods. The effects of these oscillations have been felt on coasts the world over; they have influenced coastal evolution, with repercussions on the lower courses of rivers. So an arrow leads from 'morphoclimatic systems' to the 'morphogenetic evolution' column.

The morphoclimatic systems corresponding to presentday climates, like those related to palaeoclimates, control land sculpture by permanently interfering with its general evolution; therefore arrows from the two corresponding columns end in the rectangle inscribed 'types of morphoclimatic relief'. Through the medium of lithology, morphoclimatic land sculpture depends on structure; it develops within the framework of structural relief forms and is subordinate to them. With the object of marking this difference in dimension we use in one case the expression 'structural relief' and in the other 'morphoclimatic land sculpture'. For example, in scarplands, a form of structural relief, the alternation of temperate and cold climates—such as Lorraine has experienced for example—only cause minor modifications, which result from changes in the processes of slope formation. The clifflike scarp of hard limestone becomes blunted during the Ice Ages under the attack of gelifraction. At the same time the marls of the lower scarp slope are affected by solifluction so as to produce a long and concave slope. During a warm period, on the other hand, landslips create bulges on this slope, and undermine the limestone which resumes its cliff-like form. Structural forms thus provide the framework within which the morphoclimatic processes can work. A contrary action, though small, is also possible. The behaviour of rocks is not the same in all morphoclimatic systems. Certain granites, for example, though resistant to mechanical erosion, are a prey to chemical weathering. They are much more easily disintegrated in a hot wet climate than in a dry temperate or cold one. The forms of differential erosion and denudation are thus affected, and through them the sculpturing of structural forms controlled by lithology.

The mechanisms of causality in geomorphology are thus extremely complex. One action almost invariably precipitates another action, which may in some cases be in opposition. So long as the opposing forces remain unequal there will be evolution, with the greater force dominating. If the inequality of the opposing forces diminishes, the rate of evolution will slow down, and the relation between the two forces may even be reversed. For example, during the Quaternary, we notice a general tendency for the downcutting of river valleys in many regions; but this tendency was reversed at certain times under the influence of palaeoclimatic modifications, being replaced by an opposite tendency, the rivers becoming incapable of transporting all the material derived from the slopes. The alternation of periods of downcutting and periods of alluviation, created a series of terraces separated by incisions, depending on the relative importance of the two processes. So long as evolution proceeds in the same direction and at a more or less constant speed, we have 'dynamic equilibrium'— to use a phrase coined by H. Baulig. This moving equilibrium is equivalent to the 'climax' of the biogeographers.

In other cases, the induced actions lead in the same direction as that which caused them. Thus they act as a reinforcement, producing what doctors call a system of auto-excitation; evolution accelerates by a chain reaction to attain a maximum velocity. Other forces are then brought into play, which, acting in the opposite sense, provide a brake—until 'moving equilibrium' is restored. This racing of the engine, to change the metaphor, corresponds to a more or less permanent break in the equilibrium. It may have varied causation. One of the most usual is a modification of the vegetative cover. This may have much more startling effects than tectonic phenomena which act more slowly and with more lasting effect. Man's destruction of the natural vegetation has produced a real morphogenetic crisis. Sharp climatic oscillations such as occurred in the Quaternary may have similar though less abrupt effects through the modification of the vegetation cover, especially in those marginal areas where the cover is most tenuous.

The nature of the phenomena we have tried to elucidate controls the plan of this work. It is divided into three parts, the first devoted to structural influences, the second to the general processes of geomorphological evolution and the third to morphoclimatic influences. None of these themes is independent of the others, and for this reason the overall view must never be lost.

Summary of the history of geomorphology

Geomorphological processes are difficult to observe directly because of their slow rate of operation, so they were late in being appreciated. Only a few quite exceptional observers, such as Leonardo da Vinci and Bernard Palissy, recognised them as early as the Renaissance. Leonardo, who was an engineer familiar with canal construction, was convinced that rivers erode their beds and that valleys resulted from this slow excavation. But his ideas, inscribed in notebooks, remained unpublished until the end of the eighteenth century; they were not communicated to anyone and had no influence on the development of scientific thought.

When, during the eighteenth century, the spirit of modern science began to appear, it was mainly engineers who became aware of the dynamic aspect of landforms, for they were confronted by it in their road building, canal construction and port works. Thus Lamblardie attempted to measure the drift of pebbles along the coast of the Pays de Caux, in order to estimate the amount of dredging necessary to maintain the small ports. Du Boys, a road and bridge engineer, showed that a relationship existed between the size of the particles carried by a river and the velocity of the current.

These practical men approached geomorphology with a great breadth of view, by the examination of facts which though small in themselves were of great practical importance. Even when Surrell, who followed them at the end of Louis-Philippe's reign, propounded his theory of torrents, he was only concerned with such detail as one can see on a map of 1 : 25 000 or 1 : 50 000 scale. At this level we are concerned with those expressions of the external forces that are most easily seen, take effect most rapidly, and have some immediate practical interest. This is only one aspect of geomorphology.

Outside the limited circle of technicians, forced by nature to understand her laws in order to fashion their various works, there was as yet no science of geomorphology. The manifestations of morphogenesis were too obscure to be a matter of common sense. The very soil under our feet is a symbol of stability and of permanence, and there was nothing to do but describe it. The geographer, like the historian, could only be a scholar; his role was to know all the rivers, mountains, passes, and to put them on maps. He compiled travelogues and filled in the gaps from his imagination—like the lakes in the unknown heart of Africa.

It was through geology that geomorphology had its birth, a century after the early geomorphological work of the engineers. Early geologists like de Saussure, at the end of the eighteenth century, had doubts about the stability of the earth beneath us, as they looked at fossil shells on Alpine summits. But the Flood could explain everything, and Voltaire came to the rescue of the old beliefs. However, geological evidence of geomorphological evolution accumulated with progress in the earth sciences. The Scotsman Hutton was one of the first to express it clearly. The recognition of unconformities in the English coal basins became of great economic importance. From it followed the realisation that mountains had been formed and totally destroyed by erosion before the deposition of subsequent strata. Charles Lyell, whose celebrated *Principles of Geology* (1830–33) went through many editions, and in its French translation stimulated research in that country as well, devoted himself to an analysis of the agencies which disintegrate rocks, remove their debris by running water, and in the course of geological time destroy mountains and lay down sediments from which the next generation of mountains will be built.

This academic geomorphology was thus founded on geology. It arose from the questionings of geologists and the need to find answers to their problems. From the first it had a proper scale of values—it talked in millions of years, and in terms of mountain ranges and great structural units such as the Hercynian massifs of England. It had nothing to do with the engineers, who were working on an entirely different scale. So, in the nature of things, there was a great danger of mutual ignorance in the two fields from which geomorphology developed. The great Lyell did attempt to bridge the gap by describing in detail the fissuring and fragmentation of rocks by frost action and the removal of debris by streams swollen with floodwater. But the difference in scale between these 'minor' phenomena and the building of great mountain chains made their integration in geological syntheses difficult. Here lay the basis of geomorphological error.

The more or less complete state of ignorance of the processes by which the surface configuration of the land develops led to the filling of the gaps by imagination and inspired guesswork. The popular mind had always been impressed by two manifestations, volcanic eruptions on the one hand and the battering of the coast by storm waves on the other. These two things became the basis of scientific theories—with, of course, the Flood, which at least bore the respectability of biblical theology. In the mid-nineteenth century, geology became violently split between the Plutonians who saw everything as a consequence of the fiery interior of the earth (with Elie de Beaumont's theory of crustal contraction as one aspect of this) and the Neptunians, for whom everything derived from water, sedimentation and planation. For the latter, the force of the waves was all-powerful in reducing entire continents to abrasion platforms. After all, were not such abrasion surfaces visible in the great post-Hercynian unconformities, almost always covered by marine sediments?

Fortunately, certain scientists of a more methodical and precise turn of mind stood aloof from these sterile controversies. In Germany, von Richthofen carefully observed the processes of land sculpting during his travels, and particularly in China where he noted the role of wind in the formation of loess. In France, the geologist de Margerie and a surveyor, de la Noé, joined forces in an analysis of the relief of the Jura and its relation to the disposition of the geological strata. This was the first essay in modern structural geomorphology, and the starting point for the terminology that is still in use. It occupied an intermediate position between the broad and the narrow lines on which the subject was founded.

Towards the end of the nineteenth century an American, W. M. Davis, who began as a climatologist and had a flair for systematisation, essayed the formulation of a theory of the evolution of landforms that was destined to integrate the views of the geologists. He made good use of the remarkable observations made in western United States by a whole line of scientific explorers, notably Powell and Gilbert, who had shown the great importance of water erosion in these arid lands and the influence of structure on relief. The personal contribution of Davis to the accumulation of data was virtually nil; but, as he himself admitted, he used his imaginative gifts to construct a system.

Davis viewed the problem in a geological context. It was a matter especially of explaining the unconformities between eroded folds and their sedimentary overlays. His theory envisaged a 'cycle of erosion' which may be briefly summarised thus:

1. mountains are formed as a result of sudden, even catastrophic, earth movements, as some geologists continue to aver;
2. erosion attacks these mountains, when they are formed, and generally produces a dissected relief in accordance with the already determined structures;
3. this dissected surface develops through regular and unalterable stages— youth, with rivers cutting rapidly downwards, maturity during which the downcutting slackens and the slopes are moulded and smoothed, and old age during which the rivers slow down and cease to erode, while the slopes become finally smoothed out into a peneplain, which is the evolutionary conclusion;
4. new earth movements—the 'global revolutions' of Cuvier—can start the whole process off again; the relief is rejuvenated and through the stages of youth, maturity and old age a new and final peneplain may be produced.

We are dealing with a *cycle*, because the whole process ends up at the point from which it began. The teleological spirit of the author prompted him to apply to this cycle a terminology derived from living organisms. His scheme satisfied the demands of the geologists, for it gave a clear and simple explanation of unconformities. But the scheme came too late; it was based on the geological ideas of 1830–50. It is unacceptable to geologists now because fresh data have altered their viewpoints; but it was almost outmoded before it was developed, for even then many geologists were already of the opinion that mountain-building movements are not short and sharp episodes but prolonged manifestations staggered throughout geological time. If this is so, the very foundations of the cycle of erosion crumble. There is no longer tectogenesis followed by morphogenesis, but continuous interaction between tectonic movements and the agencies of erosion. Moreover, as the earth evolves, there are no cycles but only oscillations to one side and the other of the general line of evolution. A mountain range eroded to its very roots no longer behaves, tectonically, like recent sediments. The geosyncline stabilises itself and gives place to a stable block, which is deformed in a different way; subject it to violent stress, and it cracks instead of folding. On the geological time scale, which applies to the cycle of erosion, climate, vegetation and the processes of erosion may all change. And there is no going back. The morphogenesis of the Quaternary, with its cold climates, its climatic instability, its continual oscillations in sea level arising from the glaciations, is not the same as that of the Eocene or even the Pliocene.

The cycle must thus be abandoned and replaced by an evolutionary theory, with no return to the starting point, but at the same time no direct line of progress. There are departures of greater or less amplitude from the main direction, which is ordained by the evolution of the materials making up the globe (radioactive disintegration providing the energy source), the evolution of life (especially of the vegetative cover which exercises a notable influence on erosional processes), and cosmic evolution (climatic changes, displacements of the polar axis, and perhaps variations in the rotational speed of the earth). Nevertheless, certain useful elements are worth retaining from the concept of the Davisian cycle. The author himself classified all the landforms by reference to the trilogy *structure, process, stage*. If we withdraw the cyclic element involved in *stage*, and look upon it simply as a step in an evolution which does not lead back to the starting point, and if we add to this evolution, and also to the *process* a morphoclimatic element, then the trilogy is still a worthwhile concept. Indeed it has been revived in our scheme, in a slightly modified form, at a time when the idea of the cycle has been abandoned and replaced by general morphogenetic evolution. Davis's mistake was to regard only the *stage* as cyclic; he completely ignored the possibility of changes in *process*.

The Davisian theory of the cycle of erosion was completed by that of 'normal erosion', which regards the whole evolution of relief, within the framework of the cyclic concept, as due to the work of running water. Davis erected this theory on the foundation of the excellent contemporary studies of the semi-arid regions of western USA, where indeed this is the case. But he was gravely in error in using a uniformitarian approach to apply the same principles to humid areas. The planations produced by river erosion in dry environments are not *peneplains* but *pediplains* which have a different form. They do not result from the progressive smoothing of the slopes but from the gradual extension of *glacis*, in the form of inclined planes, subject to intermittent stream erosion, which surround the residual masses, or *inselbergs*, whose steep slopes do not get smoothed. The mechanism of the evolution is totally different from that in humid regions.

The theory of normal erosion is integrated with that of the erosion cycle. It rounds it off by providing an explanatory framework for the evolution of relief. But this provoked the first unfavourable reactions, not long after its publication. Geomorphologists in Germany especially, but also in Russia, whose background was in natural science rather than (as in France) in history, began to rebel against it. They were great travellers, and even before the end of the nineteenth century they had noted the great differences in relief forms to be found in the various climatic zones. The Germans in particular were very much aware of the peculiarities of the deserts and tropical zones. Bornhardt, for example, in South Africa, had already studied the rock domes that he called *inselbergs*. Passarge had described the arid landforms of the Kalahari. Dokuchaev had founded modern pedology by demonstrating the relation between soils and climate and the part played by climate in morphogenesis. Lozinski, a Pole, put forward the idea of periglacial landforms a few years later. These ideas and methods made the scientists of central and eastern Europe implacably opposed to the Davisian concept. There were too many objections to 'normal erosion' for the theories as a whole to be accepted. Davis himself staged a fighting retreat by recognising that 'abnormal erosion' might exist side by side with 'normal erosion'. Alongside his cycle resulting from running water he erected a glacial cycle and an aeolian cycle—for at this time, notably under the influence of the German Walter, there was much overemphasis of the influence of wind erosion in deserts.

Thanks to their background of natural science, the geomorphologists of Germany, Sweden, Poland and Russia, who declined to accept the Davisian theories, provided the background of modern geomorphology. A. Penck, at the end of the nineteenth century, began his palaeogeographical reconstructions of the Quaternary glaciation of the Alps, using for this purpose a method of dating the moraines by the mode and extent of their alteration; his results are still valid. Many others, basing their studies on detrital deposits, were able to identify the various morphoclimatic phases of the past and to pronounce on their role in the evolution of the present landforms; they also studied actual processes in making their reconstructions. Högbom, in Sweden, studied the role of frost in cold climates, and in 1916 Salomon demonstrated the influence of periglacial conditions on the relief of the old massifs of Germany. Passarge even published a geomorphological map (1914) on a scale of 1 : 25 000, based on his own fieldwork. W. Penck coined the term 'correlative sediments' for the detrital deposits which enable the erosional story which they reveal to be deciphered. Whereas, in the Davisian system, only one aspect of the matter was considered, namely erosion, the accent was now on the dynamic relationship of ablation and accumulation within the same morphogenetic

20

system. The term *erosional system* must be used exclusively for the combination of transport processes that result in the removal of debris from its point of origin. Its counterpart is *accumulation system*. Unfortunately some authors still bemused by the Davisian vocabulary, use 'erosional system' in the sense of 'morphogenetic system', and this hinders the understanding of the phenomena.

Although the German and Slav geomorphologists paved the way, before the First World War, for modern geomorphology, its development was not without its difficulties. In France the work of Baulig, a faithful admirer of Davis, was partly sterilised by his adherence to the Davisian system and his postulation of tectonic stability since the mid-Tertiary, and his consequential adoption of eustatism, or oscillations of sea level, as the sole cause of subsequent morphogenesis. Certain of his suggestions are indeed astonishing, such as the invasion of a large part of the Paris Basin by a Pliocene sea—which has left no traces! In Germany, W. Penck followed the Davisian line in his theory of the *Piedmonttreppen*, or Piedmont staircase, in which, under the influence of an intermittent uplift, the steps of the staircase were carved in the landscape in between each pair of vertical movements. Lester King, in South Africa, after the Second World War, uses a similar line of thought in his theory of the pediplanation surfaces he professes to see, at exactly the same height and of late Jurassic age, in Africa, South America and Australia. All this sort of thing, it seems to the present writer, derives from a lack of real scientific approach, and it need not detain us except to put us on our guard. It stands condemned by the very complexity of the chains of causality which operate in geomorphology.

The mass of new geological observations on tectonic phenomena, and especially their evolution through geological time, together with the body of data in dynamic geomorphology on the mechanisms of land sculpture, and the new facts produced by climatic geomorphology, have enabled us the better to grasp the dialectical quality of the subject. Surface configuration, processes and deposits are inseparable. They form a dynamic system, influenced by climate, placed in a structural framework and controlled by a general evolution in which tectodynamism also plays a part. A landscape can only be understood if it is dated and correctly placed in an evolutionary sequence which is not cyclic. It cannot be interpreted unless it is identified in its dynamic context, and such identification depends in large measure on the study of the correlative deposits.

To avoid begging the question, we must thus come face to face with a series of facts of diverse nature, and thus endure the mutual control of data and interpretation. Such control is most commonly established between the surface shapes and the materials that cover them in the form of detrital deposits, alteration products and soils.

The methodology of modern geomorphology is thus founded on the analysis of process (dynamic geomorphology), on the study of correlative deposits and on dating techniques that are becoming more and more refined; these things enable us really to reconstruct the evolution of the landscape instead of imagining it. But as the mechanisms vary with the scale of the landforms considered, the techniques employed must also vary within the general framework of investigation.

Thus scientific geomorphology has only really blossomed within the last decade or two, after the Davisian heresies, in understanding the logical nature of land sculpture and the scale variations of the mechanisms involved. It has now become possible to understand the real relationships between phenomena that were formerly studied separately. The chain of geomorphological facts is now a series of continuous links, from the structure of the earth and the distribution of the ocean basins and continental areas to the sculpture of solution hollows on a rock face or the disintegration of granite. The various factors bear different weights according to the scale of the features under consideration, and the relations between them are not always the same. The pitting of rock surfaces owes more to exposure than to tectonics. The ocean basins, on the contrary, reflect the structure of the earth's crust. Geomorphological reasoning must thus be very flexible to embrace such a wide range of contexts. This explains why systematisation of the Davisian type put such a brake on progress and retarded the development of the discipline by putting blinkers on observation. We must above all things observe and reason objectively and take great care always to compare our results with those of other scientific disciplines, for nature herself is a unified whole.

STRUCTURAL GEOMORPHOLOGY

It is the internal forces that create the inequalities in the earth's surface which permit the forces of gravity to be exercised. It is also thanks to these forces that the earth's surface is differentiated into continents and oceans. Without their intervention, the ocean basins would be uniformly covered by water to a depth of over 2 600 m; despite the influence of tides and of convection currents arising from unequal solar heating, morphogenesis would be almost at a standstill. Tectonic movements, and locally vulcanism, are thus in practice the cause of all relief and morphogenetic evolution, as Davis originally observed. For this reason it is appropriate to begin our geomorphological studies by an examination of the structural factors.

In this book we consider erosion in general, without defining the mechanisms or analysing the processes, simply in order to understand its interaction with earth movements and to show how structure influences relief. Such treatment is obviously inadequate; but structure is a framework that imposes itself on morphoclimatic systems. As indicated above, the processes of dissection, whatever the climate, are influenced by the nature and disposition of the rocks, and by the general tectonic evolution of any given region. Morphoclimatic erosion is subordinate to relief produced by structure, and this subordination is partly a matter of scale. Thus in the countryside of Lorraine, an aerial view will reveal first the cuestas; and these are clearly shown on maps of 1 : 100 000 or even 1 : 200 000 scale. But, in traversing the terrain, one can see the landslips in the Oxford Clay, the scattering of limestone rubble in the hollows, the tendency for the incision of streams into periglacial embankments, and so on—in other words, the morphoclimatic landforms which link the present landscape with that inherited from the Ice Age. These are just about perceptible on a 1 : 20 000 map, but a map on this scale would hardly permit the recognition of landslips.

In general, it may be said that structural influences predominate when an area is viewed on a small scale, and morphoclimatic influences when it is seen on a larger scale. We therefore proceed from the greater to the smaller features, starting with the general configuration of the globe and the differentiation of continents and ocean basins, continuing with the largest structures like geosynclinals and platforms and ending with faults and vulcanism.

1
The globe—continents and oceans

The first and most important fact that the geomorphologist has to explain is the differentiation between dry land and the oceans; for morphogenesis beneath the atmosphere and beneath the sea obeys different laws. The structural limits of continental and oceanic areas, however, are not coincident with the coastlines, which separate the two different domains of land and water. The continents extend to a greater or less extent under the sea, forming *continental shelves*. There are thus two problems: that of the difference between continents and ocean basins on the one hand, and the encroachment of the seas on the continental shelves, on the other. We therefore study first the distribution of land and sea, then the structure of the continents and finally the fluctuations of sea level.

Distribution of land and sea

The distribution of land and sea and the configuration of the continents has exercised the attention of geographers for a long time and has been the subject of many statistical analyses. The abundance of statistical data in this field is in marked contrast to their rarity in other branches of geomorphology. The general form of the earth has been studied especially by geophysicists, for its explanation lies in the very constitution of the planet. Wegener, for example, the author of a theory of the origin of continents, one of the initial objects of which was to explain the 'fit' of the Atlantic coasts of South America and Africa, was a geophysicist. It is easy to understand, therefore, why this chapter of geomorphology has been approached in a spirit more mathematically oriented than others. Paradoxically, these continents, so limited in number, have been subject to much more study by methods of mathematical analysis than objects such as slopes, which are present in vastly greater numbers and so more apt for mathematical treatment. But this paradox is a consequence of the slow advance of geomorphological method.

The distribution of land and sea can be considered in three dimensions; for clarity of treatment we study it first in plan and then in terms of altitudes and depths.

The pattern of land and sea

Some 71 per cent of the earth's surface is occupied by sea and 29 per cent by land; the proportion of land is 39·4 per cent in the northern hemisphere against 18·6 per cent in the southern. An arbitrary hemisphere with its centre near the mouth of the Loire would include the maximum percentage of land that it is possible to get, namely 47 per cent. We might call this the continental hemisphere. The other hemisphere has only 11 per cent of land and is mainly occupied by the Pacific Ocean.

In detail, the distribution is more complex. Although the seas are all intercommunicating (except for such inland seas as the Caspian and Aral, which are few and of relatively small extent), the lands are separated from each other, a fact which has important biogeographical consequences. Moreover, the individual land masses vary greatly in size, from the 54 million sq km of Eurasia to islands that are much smaller than one square kilometre. Such differences oblige us to base an analysis on dimensions and to study separately the continents and islands.

THE CONTINENTS

The main part of the dry land is made up of the continents. If we regard Greenland as a continent, the islands only occupy 5 per cent of the total land surface. This position of Greenland is justified in the following table showing the areas of land masses in decreasing order:

Table 1.1. *Land masses* (sq km)

Asia and Europe and adjacent islands	54 200 000
Africa	29 800 000
North America *including* Greenland	24 200 000
South America	18 000 000
Antarctica	13 100 000
Australia	7 700 000
Greenland	2 175 000
New Guinea	820 000
Borneo	744 000
Madagascar	594 000
Baffin	476 000
Sumatra	473 000

Although Greenland is intermediate in size between Australia, which is certainly a continent, and New Guinea, which is not—it is 3·5 times smaller than Australia and 2·7 times larger than New Guinea—the statistical distribution changes sharply below the position of Greenland, and we are left with a series of large islands.

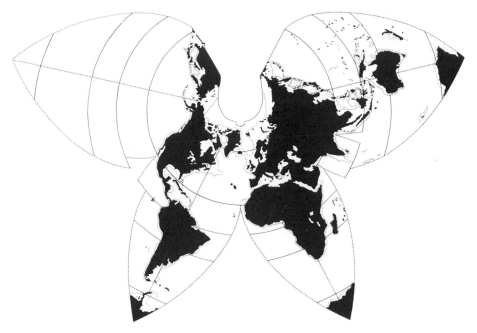

FIG. 1.1. Coherence of the areas of continental character
(*after L. Egyed, 1957, p. 107*)

The localisation and shape of the continents obey certain rules, first formulated by Chevalier in 1962:

1. Distribution is asymmetrical about the equator. Thus between latitudes 60° and 70°N land covers 70·3 per cent of the total area, whereas between 50° and 60°S it covers but 0·8 per cent. These are extreme values, but they are significant. The opposition of the hemispheres is also found at the poles, though paradoxically the north pole, within the continental hemisphere, is under the sea, whilst the south pole is on land.
2. There is an approximate symmetry of the main continental masses about a meridional plane passing along longitude 20°E and 160°W. This line crosses Europe, Africa and the Pacific Ocean. Correlatively, a plane perpendicular to this, corresponding to the meridian of 110°E and 70°W, separates a hemisphere essentially oceanic, made up mostly of the Pacific, and a land hemisphere containing Africa, Europe, the greater part of Asia and almost all of South America.
3. Outside Antarctica, isolated around the south pole, the continents are arranged in north–south wedges separated by wedges of ocean. Each continental wedge is made up of two continents: Europe and Africa, North and South America, Asia and Australia. Only Greenland stands apart—but as we have seen, it occupies an intermediate position in the size

27

classification of land masses. Each of the three double continents points southward.

4. Each of the double continents has a slight twist, with a displacement towards the west in the northern continent and a displacement eastwards in the southern one. The angles of torsion, it is true, are variable, only 20° for Europe and Africa, but 40° in the case of the Americas and 55° for Asia–Australia. Naturally, the oceans reflect this same torsion, which is particularly noticeable in the case of the Atlantic.

ISLANDS AND SEAS

Within the interior of continents there exist certain enclosed seas, such as the Caspian and Aral. Their waters are salty, and this may justify their being called seas, but such an appellation is merely traditional and has no geomorphological justification. These seas owe their existence to a semi-arid climate that produces sufficient evaporation to render them salt and prevent their outflow by rivers to the real sea. If the climate were more humid, they would have become lakes, like the Great Lakes of North America, Lake Baikal or the lakes of East Africa.

Similarly, there is a fundamental difference between the continents, which are fringed with islands and contain no real seas, and the oceans in which one constantly finds islands even in the middle of the most vast water expanses, such as Hawaii, Easter Island, Ascension, St Helena, Kerguelen and St Paul.

A study of the distribution of islands must recognise two characteristics that are significant:

1. The dimensions of islands are extremely variable. A whole series of them (actually 31 per cent of the total) exceed 25 000 sq km. Many others are less than 2 500 sq km. The majority of the large islands are less than 200 km from continents, whereas the smaller islands lie much farther away (e.g. Oceania, West Indies, the South Shetland arc and the islands of the South Atlantic). Of the larger islands located more than 200 km from a continent, most are part of island groups such as the West Indies and the East Indies. Only New Zealand, Madagascar and Spitzbergen are both isolated and located more than 400 km from a continent.

2. Their distribution also shows great differences; three main types may be recognised: (*a*) Islands, often quite large, situated close to continents, and having a geological structure similar to that of the neighbouring land mass, with the same geological formations and tectonic features. The best example is provided by the British Isles, but the north-Canadian archipelago is similar, and although at a greater distance, Spitzbergen and Borneo fall into the same class. (*b*) Isles of various sizes, disposed like garlands and of peculiar geological constitution. Such are the Japanese archipelago, with its volcanic rocks completely different from those of continental China, Formosa and its neighbouring archipelagoes, the lesser

Antilles, the Kuriles and the Aleutians. Volcanic rocks are of frequent occurrence in this type, even though the surface may be made of coral; many atolls have a volcanic foundation. (*c*) Islands isolated in the midst of oceans, far from any land. These, the 'oceanic islands', are always volcanic and nearly always small, like the south Atlantic isles, Madeira, Easter Island, Hawaii and Kerguelen. Iceland might be included in this category, in which it would provide the largest example (103 000 sq km) if it were not too close (220 km) to Greenland.

Dimensions and distribution thus provide a means of classification. Almost all islands fall into one of these three categories: oceanic islands, islands arcs, and subcontinental islands close to the mainland and with similar structure. Most of the subcontinental islands are separated from continents by depths of less than 200 m and rest on the continental shelf. Some islands can only with some difficulty be fitted into this classification, but they are few in number. New Zealand, for example, might be regarded as of the arc or garland type; it is elongated in shape and is prolonged by further islands, and it has a structure involving geosynclinal sediments and volcanoes. Spitzbergen and Madagascar are really subcontinental islands despite their distance from the nearest mainland (400 and 420 km respectively) and the depth of the separating seas, mainly by reason of their structure.

The most significant fact is the peculiar geological nature of the islands that lie far from continents. They are as different from the continental land masses as lakes are from seas. And like lakes, moreover, they occupy but a minute part of the earth's surface. They are, indeed, anomalies. The same result accrues from an analysis of altitudes.

Distribution of heights and depths

The statistical distribution of heights and depths can be portrayed graphically in a hypsographic curve.

The hypsographic curve

This shows the area occupied on the earth's surface by the different slices of submarine depth and terrestrial altitude. The development of echo-sounding techniques within the last few decades has enabled a satisfactory precision to be given to oceanic depths. On the other hand, certain doubts still persist regarding the continents, by reason of the existence of the Greenland and Antarctic ice caps. It is easy enough, in the light of modern exploration, to draw contours of the ice surface, but this means little for the accumulation of ice is the result of climate, not structure. More interesting is the altitude of the subglacial rock surface, and this we are beginning to know as the result of geophysical surveys. Unfortunately this subglacial surface altitude is a false one since the crust is depressed by the

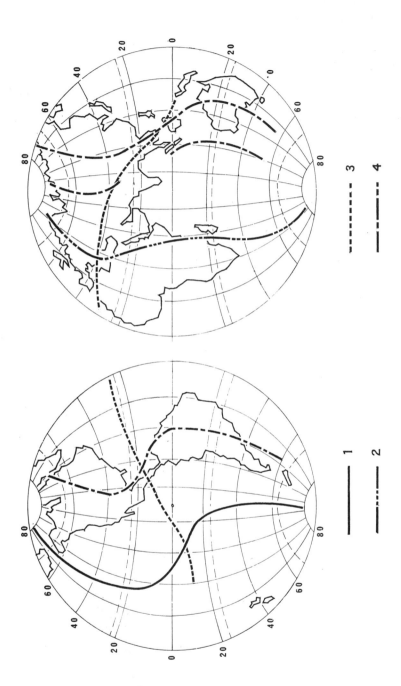

Fig. 1.2. Sigmoid axes of continents and oceans (*after P. Fourmarier, 1962, p. 1416*)
1. Pacific axis   2. Eurafrican axis   3. Antillo–Himalayan axis   4. American and Asiatic axes

weight of the ice. The most satisfactory method would thus consist, despite the possibility of error, of calculating the altitude that the rock surface would attain if the ice were removed and isostatic recovery took place. The considerable extent of Antarctica (13·1 million sq km of which 12 million are ice covered) makes this problem a matter of some importance, for the continent represents 8 per cent of the total land surface.

A study of the hypsographic curve leads to the following conclusions:

1. The irregularities of the land surface are, all told, features of quite minor importance. The radius of the earth is 6 378 km at the equator and 6 356 km at the pole; the difference is 21 476 m. But the highest mountain, Everest, is 8 848 m and the deepest ocean trench, off the Philippines, descends to 11 516 m; and the difference between these two extremes is 20 364 m. Thus the maximum relief on the earth's surface is of the same order of magnitude as the difference between the polar and equatorial radii. Perhaps this corresponds to a limiting value arising from the very constitution of the earth and its movement in space? In any case, the irregularities in the earth's surface, its relief, are relatively slight, and only reach 0·34 per cent of the radius of the globe.

2. The extremities of the hypsographic curve are of the same order of magnitude. The greatest marine depths are between 8 000 and 11 516 m and the greatest heights lie between 6 000 and 8 848 m. However, there is a skew in favour of the depths. A dozen marine trenches have maximum depths greater than the altitude of Everest. Their median is 10 210 m whereas that of the ten highest mountains is only 7 000 m if one takes mountains from different areas. The difference is about 3 600 m. It would appear that the great depths are more easily maintained than great altitudes. The contours of the great depths are noticeably more extensive than those of the highest mountains.

Nevertheless, the hypsographic curve has a generally sigmoidal form, which shows the influence of Gauss's law on the distribution of heights and depths.

3. The hypsographic curve departs appreciably from a true sigmoid, however, through the existence of two flats, which indicate an influence contrary to a true Gaussian distribution, that can only reflect the structure of the earth's surface.

The altitudinal distribution according to Cailleux (1965) is shown in Table 1.2.

One of the flats is found at altitudes between − 1 000 and + 1 000 m, or more exactly, between − 200 and + 1 000 m. It corresponds to the greater part of the land surface and the *continental shelves*. The second flat is found in the oceanic depths between − 3 000 and − 6 000 m, in particular between − 4 000 and − 5 000 m. This corresponds to the vast submarine expanses with little relief, the *abyssal plains*.

These two flattenings in the curve are separated by steeper slopes,

Table 1.2. *Altitudinal distribution*

SLICE (metres)	AREA OF EACH SLICE (per cent)		CUMULATIVE AREAS	
	A	B	A	B
Above 5 000	0·1	0·2	0·1	0·1
4 000–5 000	0·2	0·2	0·3	0·3
3 000–4 000	0·4	0·4	0·7	0·7
2 000–3 000	2·7	1·4	3·4	2·1
1 000–2 000	4·5	4·3	7·9	6·4
500–1 000 ⎫			13·5	12·3
300– 500 ⎬	20·6	22·1	19·0	18·6
0– 300 ⎭			28·7	28·7
0– −200 ⎫ −200– −1 000 ⎭	8·8		34·7 37·6	
−1 000– −2 000	3·7		41·3	
−2 000– −3 000	4·0		45·3	
−3 000– −4 000	15·3		60·6	
−4 000– −5 000	21·5		82·1	
−5 000– −6 000	15·0		97·1	
−6 000– −7 000	2·8		99·9	
Below −7 000	0·1		100·0	

A   Altitude of actual surface (or of ice caps).
B   Altitude of solid crust after isostatic recovery.

between −3 000 and −200 m, which coincide with the *continental slope*. Around most of the continental areas, beyond the edge of the continental shelf, there is a zone of much steeper slopes that leads down rapidly to the abyssal plains. The shelf itself is a platform with but little slope, and often bearing islands that have the same structure as the neighbouring continent.

The hypsographic curve records these orographical states because they are well defined and occupy sufficiently large areas of the globe between altitude limits that are but little variable. They are characteristics of major importance, resulting, as we shall see, from the very structure of the earth.

The arrangement of the major orographical units

The hypsographic curve reveals the following schematic arrangement:

1. continental areas dominated by larger plains, hills and plateaux of under 2 000 m altitude, which are extended without a break in a continental shelf covered by an epicontinental sea, of which the finest example is in northwest Europe; the Baltic, the North Sea, the English Channel, the Bay of Biscay and the Irish Sea;

2. oceanic basins characterised by vast areas between $-3\,000$ and $-6\,000$ m: the abyssal plains;

3. continental areas of much more limited extent, with altitudes of over $2\,000$ m. These are the mountain ranges, of great length, such as the Himalayas, the Andes, the Cascades and the Rockies;

4. depressions of equally limited extent in the ocean basins: these are the submarine trenches, with a depth of more than $7\,000$ m. Like the mountain chains, these are elongated and often arcuate: the principal difference is their much greater smoothness of relief.

Let us now see how these units are distributed over the earth's surface.

CONTINENTS

A statistical law was formulated by Matschinski, to the effect that the mean altitude of continents increases with their area. Thus Eurasia reaches 860 m, Africa 660 m, Australia 330 m. But these figures mask the great diversity of relief.

The principal mountain chains, in which vast areas exceed $2\,000$ m, are grouped in two great systems:

1. A mainly longitudinal system around the Pacific Ocean, with the series of ranges in western North America, enclosing high plateaux (Colorado and Central Mexico), continuing through Central America, and then the Andes. In Venezuela and Colombia a bifurcation originates, forming the discontinuous island arc of the West Indies. On the western side of the Pacific, the chains of the East Indies, the island arcs of Formosa and Japan and the Soviet Far East, constitute a more or less symmetrical system. The whole of this area is characterised by vigorous and continuing tectogenesis, which is evidenced in frequent and violent earthquakes and a large number of volcanoes, many of them still active.

2. A mainly latitudinal system, oriented east–west, in the northern hemisphere, and corresponding to the Tethys of the geologists, comprising the ranges of the Mediterranean, then those of Iran and the Himalayas. Here again a bifurcation takes place, in the neighbourhood of the Caspian, which gives rise to a series of chains from the Caucasus to Lake Baikal. The two branches enclose an enormous system of high plateaux, culminating in Tibet, which is a greater counterpart of the plateaux of western United States, Mexico and Peru–Bolivia. The greatest altitudes in this system are found in Asia, and this is responsible for the considerable mean elevation of the whole continent. There is one major difference from the circum-Pacific system, and that is the scarcity of volcanoes, which are confined to certain isolated locations as in Armenia and the Elburz.

The two systems meet in Indochina, where the Himalayan system ends against the circum-Pacific chains. In this sector the two directions interfere and lose their regularity, especially in the East Indies, with their great

33

island arcs. Both in Java and in New Guinea, one finds the Himalayan orientation resumed, though displaced far to the south. At the western end of the system there is no similar occurrence: the Mediterranean chains fray out and disappear on reaching the eastern margin of the Atlantic ocean. Nevertheless, as Cailleux remarked in 1965, the twisting already noted in the continental land masses (above p. 28) is found principally in these latitudes. The two facts are probably related to the deep structures of the earth's crust.

The great mountain systems just described are found, in the main, on the edge of continental areas. This is most obvious in North and South America, but it is well shown in Europe also, where the high and dissected relief of the Mediterranean region lies next to vast areas of lower elevation and relief. In Asia the situation is much less clear, for the highlands are more centrally disposed, around the Tibetan plateau, and extend into central and eastern Siberia. Nevertheless, orographic asymmetry is the normal case. Thus continental areas are commonly highest on their edges, close to the sea. Even Europe does not escape this characteristic, for outside the Mediterranean, the Atlantic coast is lined by mountains for much of the way, from Galicia to North Cape by way of Ireland, Scotland and Norway. In contrast, plains are dominant from the White Sea to the Pannonian Basin and from the London Basin to the Urals, with the sole and relatively slight interruption made by the Carpathians. Even in Africa the same disposition of relief is apparent, with the Guinea Highlands and the Atlas mountains, and in particular the highlands aligned from Ethiopia to the Drakensberg.

This peripheral location of mountain systems has profound geomorphological repercussions. It hinders the drainage of the central parts of continents, where lakes or basins of inland drainage occur. At the present time the only lake to exceed 100 000 sq km is the Caspian, but in the Tertiary period, and even in the Quaternary, extensive lakes were maintained for long periods in the Pannonian basin, in western USA, in central Australia and in the Congo basin. Many large rivers were endoreic, as the Volga is today, because they flowed into structural depressions in which the climate was too dry for the lake which they fed to overflow to the sea. This was the case with the Niger until the later part of the Quaternary.

Another repercussion of the position of the great mountain systems, though a more indirect one, is their effect on climatic distributions, and sometimes their interference with the planetary wind system. The regular zonation of climates in West Africa is permitted by a not-too-accentuated relief; but the mountain masses of Asia, which are in part responsible for the monsoons, give to this continent unusual climates which are very different from the normal zones. The western mountain barrier in North America is responsible for a vast extension of semi-arid climates in the west central part of the continent, while the foggy deserts so characteristic of the west coast of South America result directly from the configuration of the continent, which, tapering to a point in relatively high latitudes acts like a harbour mole in channelling the Antarctic meltwater that gives rise to the Humboldt current, while the Andean ranges isolate the littoral climates from the influences of air masses of continental or Atlantic origin.

In addition to the mountain systems, the continents are crossed by belts of faulting, or rift valley systems, which certain authors have stressed. The clearest and the most important of these is that in East Africa, which continues northwards via the Red Sea to the tectonic trough of the Dead Sea. Numerous other such systems have been suggested, but they are perhaps attempts at systematisation rather than directly observable conditions.

OCEAN BASINS

Some 58·7 per cent of the earth's surface lies between sea level and − 2 000 m. On the edge of the basins is the *continental slope*, between − 200 and − 2 000 m, which is the real edge of the continents. Here is a slope, averaging 4 or 5 degrees, which is often dissected by submarine canyons which have the appearance of river valleys. As we shall see, this slope corresponds to a difference in structure. There are three series of arguments for regarding it as the real edge of the continents: structural, geomorphological and orographic.

Beyond the floor of the continental slope, the ocean floor is much less accidented, and we reach the abyssal plains, which have very gentle slopes and very little relief; and these sink slowly to the deeps which may go down to − 7 000 m. But the areas below − 6 000 m are of small extent. The general lie of these abyssal plains is rather like that of the Russian plain with the Caspian basin.

The abyssal plains form large basins with a diameter of several thousand kilometres; they are particularly large in the Pacific and Indian oceans. Two types of relief interrupt them, the trenches and the chains of islands.

The abyssal trenches are never found in the central parts of the ocean basins, but always close to the land, whether it be the island arc (as in the case of the Kermadec and Tonga trenches alongside the New Zealand–Tonga island belt) or more rarely the actual continental edge (as off the Chilean coast). The deepest trenches are those that follow an island arc (the Philippine, Mariana, Tonga, Kurile, Japan, Kermadec, Bonin, Puerto Rico, New Hebrides and Solomon Islands trenches, are the first ten in order of depth). On the edge of these trenches there is no shelf or continental slope; the descent from the land is rapid and steep, with slopes often exceeding 10 degrees, on which submarine landslips are frequent. Like the great mountain chains, therefore, the oceanic deeps are close to a boundary. They are aligned along the edge of the ocean basins, especially the western side of the Pacific, where they are deepest and most numerous. And they are generally associated with island arcs.

Island arcs are strings of islands commonly situated on submarine ridges. Some of them are prolongations of continental mountain chains, like the

Kuriles, Aleutians and West Indies, and so isolate more or less completely certain sea basins, generally much smaller and shallower than the oceans. Thus the Bering Sea floor drops to $-3\,900$ m, that of the Sea of Okhotsk to $-3\,550$ m; another example is the Caribbean, which is deeper and is sub-divided into two basins, not including the Gulf of Mexico. In general the bed of these seas is an inclined plane, deepest at the end nearest to the land (off Northern Hokkaido in the case of Okhotsk, and at the southwest end of the Bering Sea). Some other island arcs are isolated in the middle of the oceans, such as those making up Polynesia, Micronesia or Hawaii. They are from 1 500 to 3 000–4 000 km long.

Island arcs are usually rich in volcanoes. Some of them indeed are purely volcanic, apart from the coral growths that have accumulated on them. All those that are well isolated in the middle of oceans are volcanoes, rising from water depths of between 2 000 and 4 000 m. But some of the volcanoes rising from the ocean bed do not reach the surface as islands; these are the submarine volcanoes, of perfect shape but often with truncated tops at between 1 000 and 2 000 m below sea level. These are sometimes called *guyots* after the scientist who first studied them.

Somewhat similar to an island arc is the mid-Atlantic ridge, in the shape of an S, running from Iceland to Bouvet Island and marked by occasional volcanoes (Azores, St Paul, Ascension, Tristan de Cunha, Gough and Bouvet). But this is not comparable with the others; it is much longer and wider, and is interrupted by 'cols' (e.g. the Romanche trough); further-more, islands are very rare. It would seem that the structure is different from that of the island arcs of the Pacific and Indian Oceans.

All told, then, the boundary between continents and oceans is fairly sharp. It may take the form of a continental slope or of oceanic trenches adjacent to island arcs (as in Japan, the Riu-Kiu islands and Chile). The trenches are always on the edge of major ocean basins, the mountain chains on the edge of continents. All this seems to indicate that the zone of contact between continents and oceans is a critical one, in which the earth's crust has undergone important deformations. We now turn to an examination of this crust.

Structure of the crust and origin of continents

Only the superficial layer of the crust is accessible to observation by direct inspection or through boreholes. But the deepest borehole has only reached 4 000 m, and our knowledge of the crust must be gained by indirect geophysical methods. Such methods have hitherto been the only ones available for the study of the ocean beds, until recent American and Russian attempts at ocean floor boring. We study first the crust as a whole, using the information derived from geophysical surveys, and then the structure of the continents, which we can observe more directly.

Structure of the crust

Our knowledge of the crust has been largely derived from the study of earthquakes, so we may begin by examining the kind of information that they can furnish.

DATA DERIVED FROM EARTHQUAKES

Earthquakes are the result of shocks produced within the crust, giving rise to vibratory waves. These waves are recorded by high precision pendulums, the oscillations of which are traced by a pen on a revolving drum; the apparatus is called a seismograph.

The shocks that set up the wave motion are of varied origin. Some accompany volcanic eruptions; these are of superficial origin and give no information about the crust. Others originate at much greater depths and result from the tensions to which tectonic forces subject the crust. These tensions encounter great resistance from the solid rocks, and they build up very slowly until suddenly they provoke a slight movement of two masses which are in contact along a fracture plane or fault. The surfaces being very rough, this chafing generates a vibration, which is the earthquake. This is why in general, when movement along faultlines occurs, there is an earthquake. The two phenomena are linked and indicate that the tensions have overcome the frictional resistance of the rock masses and have provoked a slight readjustment of the crust.

The waves from a seismic shock travel at different rates according to the nature of the material through which they pass, and in moving from one substance to another they are refracted like rays of light through a prism. Part of the wave may also be reflected from the zone of contact, and a single seismic shock may thus give rise to trains of different waves, some reflected and some refracted. We can thus distinguish primary, secondary and superficial waves, the last being the slowest, which arrive last. When the waves from a single shock are recorded by several seismographs at different localities on the globe, the differences in the time taken may be determined precisely, and after the manner of triangulation, the point of origin may be located not merely as to latitude and longitude but also as to its depth. This kind of work has been performed at many observatories for decades, so that a considerable body of data exists. This enables us to define seismic zones, that is areas from which the shocks originate. Naturally earthquakes are classified by their intensity or magnitude, and frequency is also taken into account. The *epicentres*, that is the points on the earth's surface situated vertically above the *source* (the actual point of origin of the waves), are plotted on maps. Sometimes artificial earthquakes have been simulated, e.g. by ordinary or atomic explosions, in order to study the structure of certain parts of the crust, as in the Alps.

An earthquake is thus characterised by its epicentre and its depth. These two dimensions enable us to learn something about the structure of

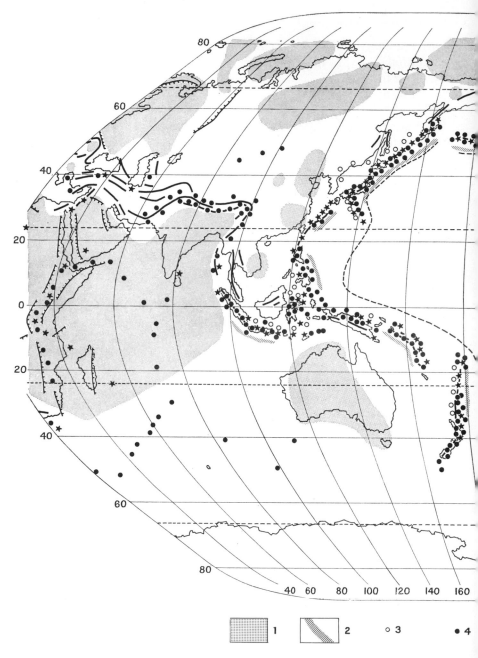

FIG. 1.3. Global distribution of volcanoes and earthquakes
1. Old 'shields', stable regions unaffected by folding since the beginning of the
    Palaeozoic
2. Oceanic trenches
3. Deep-seated earthquakes (deeper than 300 km)

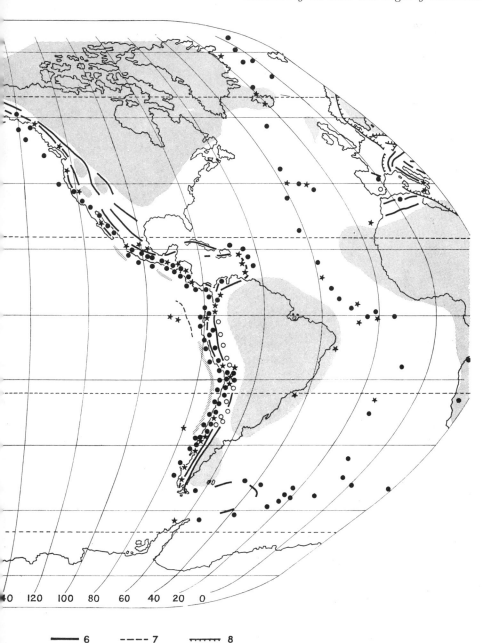

━━━ 6   ─ ─ ─ 7   ┄┄┄ 8

4. Shallow and intermediate earthquakes (less than 300 km depth)
5. Volcanoes, active during historic time
6. Axes of major Neogene (upper Tertiary) folding
7. The 'andesite line'
8. Major faults

39

the crust. But the speed of the shock waves is also of interest, for it depends on the elasticity and density of the materials traversed. We can thus get some idea of the nature of those parts of the crust that are inaccessible to direct observation. Furthermore, it is possible to survey the nature and disposition of the rocks beneath the surface without boreholes, simply by creating artificial shocks. This kind of research, however, is most easy for comparatively small depths, and geophysicists have calibrated the rates of wave propagation for all the ordinary types of rock. When it comes to the interior of the earth, matters are not so simple, for we do not know what are the properties of materials at the very high temperatures and pressures that we assume must exist at great depths. Nevertheless a certain global estimation is possible, for thanks to astronomical methods, it has been possible to calculate the average density of the earth. This works out at 5·52; as the mean density of the sedimentary rocks of the crust is 2·5 we must accept that the material of which the earth is made decreases in density from the centre towards the surface. This density, by reason of the law of universal attraction, controls the coefficient of acceleration due to gravity, g. But g varies over the surface of the globe. Where it is greatest, we speak of a *positive anomaly*, which indicates that high density material is thicker at that point, in a direction vertically downwards. Conversely, if the value of g is less than the average, we have a *negative anomaly*, which indicates a deficiency of mass, and thus a greater thickness of the lighter crustal material. One would therefore expect g to be abnormally low under the oceans, because the density of water is much less than the density of the crustal rocks; but it is not, and this implies some form of compensation which can only result from the presence of denser crustal material under the oceans.

These considerations of gravimetry help in the interpretation of seismic data, and by combining both, we can get a fair idea of the constitution of the earth.

THE CRUST AND ITS CHARACTER

The conclusions of geophysics postulate a globe formed of layers of decreasing density from centre to surface. In fact, the following are generally recognised:

1. In the centre, at a depth of more than 5 000 km, a *core* with a density of 11 to 17 or 12 to 18. It should be solid, and formed mainly of iron and nickel, like meteorites which are the debris of other planets.
2. A *shell*, between 2 900 km and 5 000 km deep, also made of iron and nickel, but more fluid, and with a lower density, estimated at between 9 and 11·5.
3. A *mantle*, formed of material rich in iron, which is the parent of certain lavas called peridotites. It lies between 40 km and 2 900 km deep and is probably solid, with a density of 3·3 to 6·7.

4. A superficial *crust*, some 40 km thick, and subdivided into an upper layer, mainly of granite with a density of 2·7, underlying the sedimentary rocks, and, below 17 km a deeper layer, equally solid, and basic, with a density of 3 and made largely of magmas resembling basalt and gabbro.

The core and the shell correspond to the *nife* of older writers, whilst the upper part of the crust is the *sial*, the granite being rich in silicates of alumina. The *sima* is the lower part of the crust. Of particular importance is the layer at a depth of 32 to 38 km, for this produces a striking modification of the seismic waves, and reflects them. This is the *Mohorovicic discontinuity*, which American authors abbreviate to *Moho*. It corresponds to the base of the crust.

Seismic shocks have their origin only in the upper 700 km of the globe, that is in the mantle and especially in the crust. They frequently originate in the vicinity of Moho, which appears to be a sensitive zone in the earth's constitution. Easily recognisable by seismology, its depth has been precisely ascertained. The figure of 32–38 km given above is a mean, but there are important variations.

The Mohorovicic discontinuity behaves as follows:

1. Beneath important mountain chains, those great mountain systems described above (p. 33), Moho lies deeper, at 45–60 km, or even perhaps 70 km. It is most depressed beneath the highest ranges, those which owe their altitude to particularly vigorous recent crustal deformation. Under these mountains there are thus considerable roots, much deeper than the corresponding elevations. The inequalities in the earth's surface thus correspond to even greater inequalities, and in an inverse sense, in the level of Moho. In other words, the great mountain chains coincide with considerable thickenings in the upper part of the earth's crust.

2. Beneath the oceans, on the contrary, Moho rises to a much lesser depth and is to be found, under the abyssal plains, at only 5–7 km beneath the ocean floor. The crust is here abnormally thin, exactly the opposite of what happens under the mountains. Its disposition is regular, like that of the submarine relief. It does not rise beneath the volcanic islands, but does plunge under the great trenches. It slopes at between 10 and 45 degrees from the oceans towards the continents where mountain ranges border the lands, and behaves similarly in relation to the island arcs.

3. Finally, beneath the continents, away from the main mountain ranges, Moho remains stable, as under the oceanic abyssal plains, but at a depth of 30–35 km. Some local anomalies have been noticed in areas of heavy faulting and vulcanism, as in East Africa.

The combined results of seismology and gravimetry enable us to appreciate the peculiar relationship of the earth's crust to the lower layers. The latter form a continuous shell around the core, whereas the upper crust, the sial, appears to be discontinuous; it is limited to the continental areas and is almost absent beneath the great ocean basins. The sima, however, though continuous, has an irregular upper surface, much more irregular than the surface of the lithosphere. Beneath the ocean floors, geophysical prospecting seems always to reveal a relatively thin skin of sediments, only a few thousand metres thick, with frequent intercalations of volcanic material, and then immediately below, a layer in which the seismic waves behave as they do in the basalts. There is thus no sial, which explains the rise of Moho under the continents; on the other hand, seismic waves travel, as far as Moho, in the same way as they do through granite. These observations lead to the following conclusions:

1. Moho is the zone of contact between the granitic upper crust, the sial, and the basaltic lower crust, the sima;
2. sial is limited to the continental areas and comprises the foundation of the continents, beneath the sedimentary and metamorphic rocks;
3. the ocean basins are characterised by the existence of a thin layer of sediments and volcanic deposits lying directly on the sima, which here closely approaches the surface of the lithosphere.

There should thus be a fundamental difference of constitution between the continental areas and the ocean basins. It was to investigate this hypothesis that 'operation Mohole' was planned by American scientists, to bore through the ocean floor as far as Moho. The programme was abandoned after a year or two of preparatory work, largely because of the colossal expenditure involved. Meanwhile, the opposition of sima and sial remains the basis of present theory relating to the dynamism of the earth's crust.

DYNAMISM OF THE CRUST; ISOSTASY

On the basis of the difference between the upper part of the crust and the immediately underlying material, the theory of isostasy has been developed. As we have seen, it is agreed that the variation in wave propagation results from the difference in density of these two media. The theory is supported by gravity measurements and by what we know of the mean density of the globe. Because of the differences in density, sial, the lighter material, floats, as it were, on the denser sima. This is the principle of isostasy. Basically it recalls the principle of Archimedes, applied on a global scale. The difference in density being but slight, about $1:10$, the sinking of sial into sima is considerable, somewhat like icebergs in the sea.

The behaviour of Moho confirms this hypothesis. Its depth of 32–38 km beneath continental areas that have not been subjected to recent folding would represent the normal degree of sinking of sial into sima, given the density difference and the mean altitude of the continents. Its greater depth under the great mountain system results from the greater weight of these localised thickenings in the upper crust. And its small depth under the ocean basins is explained by the absence of sial in these regions. The seismic activity observed in the immediate vicinity of Moho results from tensions and readjustments at the sial–sima contact. All this is a weighty argument in favour of isostasy.

There are other arguments as well; for example the localisation of earthquakes. The major seismic zones (Fig. 1.3) occupy but a tiny fraction of the earth's surface. According to Gutenberg, 80 per cent of the total energy released by earthquakes is liberated in the circum-Pacific zone, on the periphery of the ocean, especially in the volcano-studded island arcs on its western border and in the western cordillera of the Americas. Another 15 per cent emanates from the east–west mountain system of middle latitudes, from Gibraltar to Assam in the east and the arc of the West Indies in the west. The remaining 5 per cent is concentrated in certain other small areas, notably the East African rift valley system (2·3 per cent) and the middle Atlantic ridge (1·9 per cent). In contrast large areas such as the ocean basins and the continental plateaus are completely devoid of earthquakes. Earthquakes are thus localised along the great mountain systems which are still subject to orogenesis, especially where, as around the Pacific, the mountains lie at the junction of continents and oceans. In such regions sima plunges rapidly under the thickened sial, and the Mohorovicic discontinuity descends steeply under mountain chains and island arcs.

Lastly, the study of the consequences of the last (Würm or Weichsel) glaciation also contributes serious support to the theory of isostasy. Indeed it was such study that gave rise to the theory in the first place, for incontrovertible observations have been made, from the eighteenth century, of the important uplift in Scandinavia of areas from which the ice sheets had more recently retreated. Similar observations have since been made in other similar areas, notably in North America. It would appear that stable continental areas were depressed by the weight of ice that covered them, and that they recovered when the ice melted. This particular application of the theory of isostasy is known as glacio-isostasy.

Glacio-isostasy has the advantage of enabling us to analyse better the mechanism of the movements, and it pinpoints the difference of scale that separates isostasy from Archimedes's principle, despite their similar nature. Sima is not liquid, and it reacts to the imposition and release of weight slowly and after some delay. Several thousand years elapsed after the retreat of the ice before isostatic recovery set in. Whilst the ice disappeared about 8 000 years ago, and the beginning of the retreat occurred much earlier than that, the isostatic readjustment is still not complete, even though it has been going on for 2 000 years. However, a delay of 10 000 years is of no consequence in relation to geological time; the Quaternary period alone represents a million years, and geochronology takes a million years as its usual unit! This difference in time scale corresponds, as always, to a difference in space scale. Gourinard (1952) has shown that isostatic reactions can only take place within compartments that do not exceed a certain threshold dimension. In the case of well-marked divisions closely circumscribed by tectonic features, such as the basins and massifs of the

Oran Tell, this threshold is several tens of kilometres in each direction, say an area of 1 000 sq km. For less closely delimited blocks, the threshold is larger; but for areas smaller than the threshold dimension, the forces of resistance are too strong for isostatic movement to take place; the crust takes the strain without flinching.

This notion is very important for the geomorphologist. It indicates that the cutting of a valley, below the threshold dimension in area, cannot provoke crustal disequilibrium; on the contrary, general denudation resulting in dissection of a mountain mass like the Alps or the Massif Central, may pass the threshold after a certain time and lead to a reaction in the form of a general uplift to compensate the lightening of the load. But this could hardly take place save over an immense period of time, running into millions of years. In a similar way, the overloading of an area through the deposition of large quantities of detrital sediment is also able to provoke a lowering through isostatic adjustment. This is what happens in certain intermontane basins and in some piedmont regions. In both cases we are dealing with geological time. We should be gravely in error if we invoked isostasy to explain river terraces; in this case we are well below the threshold, and the area behaves in a stable fashion.

We are in possession of sufficient evidence to confirm that the theory of isostasy is true, and fits well the observed facts. It throws light on the results of gravimetric surveys and helps to interpret them. When isostatic equilibrium is reached there are no gravity anomalies. In effect, the bulges of the crust, which have an excess of mass compared with their neighbouring areas, where, at the same altitude, the air or the sea have a much lower density, are compensated by the penetration, at the base of the upper crust, of the less dense sial into the denser sima. The density difference being slight, the roots of the sial must penetrate more deeply into the sima for compensation to take place and equilibrium to be restored. Beneath the ocean basins, the uprising of dense sima compensates the absence of sial and the depression in the surface of the lithosphere. But isostatic equilibrium is not always realised; certain areas are characterised by a positive gravity anomaly. This implies that in order to reach equilibrium, the sial must thicken, which in turn demands the formation of a mountain range since such a thickening of necessity takes place both downwards and upwards at the same time. Conversely a negative anomaly tends to disappear through the rise of sima and the thinning of sial, a process which can be realised at the surface of the lithosphere through denudation. These are but long-term tendencies, however. They are not often clearly apparent in the results of gravity surveys, for they may be obscured by anomalies of a different origin, such as those due to the nature of rocks and by short-term disequilibrium in the opposite direction. The interpretation of gravity surveys is always fraught with difficulty.

Isostasy has been used as the starting point for the development of other theories, of which the most notable is Wegener's *continental drift*. According to this hypothesis, continents, like rafts of sial floating on sima, have been displaced, in part by centrifugal force drawing them towards the equator and in part by tidal effects that push them towards the west. The Atlantic on this theory, would be simply a great rift, gradually enlarged, that ended up by separating the Americas from Eur-Africa, opening progressively from south to north. The telescoping of Africa and Europe would produce a clamping effect that gave rise to the Alpine–Mediterranean mountain systems. The Himalayas similarly resulted from the squeezing together of the Deccan and Tibet. These ideas were developed some forty years ago by the Swiss geologist Argand. Wegener himself emphasised particularly the shape of the two sides of the Atlantic and showed how they fitted. He attributed the rise of the Rocky–Andes system to the resistance encountered in the westward movement of the American continent. Conversely, the migration of Asia accounts for the abandonment of detached fragments which now form the island arcs on the western border of the Pacific Ocean. The geomorphologist L. C. King has even claimed to have discovered at similar altitudes an ancient erosion surface, of late Jurassic date, which antedates the separation of Africa and South America!

There are still some geomorphologists who entertain considerable doubts about Wegener's continental drift and the tectonic explanations of Argand. As for King's ideas, the present author has found nothing to support them, either in Brazil or in West Africa. At one stage, the continental drift theory experienced a series of important setbacks. Cailleux, for example, has remarked that it is indeed strange that the continental fission should have been delayed until the end of the Jurassic, when the earth's crust was already effectively consolidated, and that if we must accept the idea of drift, then it ought to have occurred in the pre-Cambrian, 2 000 million years earlier, when the crust was less well consolidated. He has also pointed to the fact that the substratum of the continental blocks is always formed of folded rocks, in which the folds are often much more severe than those to be seen in recent mountains. Since this folding must have occurred before the drift, it is difficult to explain it. Finally there are palaeogeographical arguments; Australian geologists, for example, have shown that there has been scarcely any modification of the western coast of their continent since the Palaeozoic. Others have demonstrated that the Mozambique Channel has been in existence at least since the Trias. It is then impossible to reunite South Africa, the Deccan, Madagascar and Australia, let alone Brazil, into the Gondwanaland continent that was allegedly disintegrated during the Mesozoic era. Furthermore, despite the precision of the techniques, no measurements yet made have shown the slightest signs of change in the relative positions of Europe and North America, which should be occurring if the two continents are drifting apart as postulated by Wegener.

The data accumulated by various workers have been brought together by Cailleux to form a new theory which seems more satisfactory. The sial

45

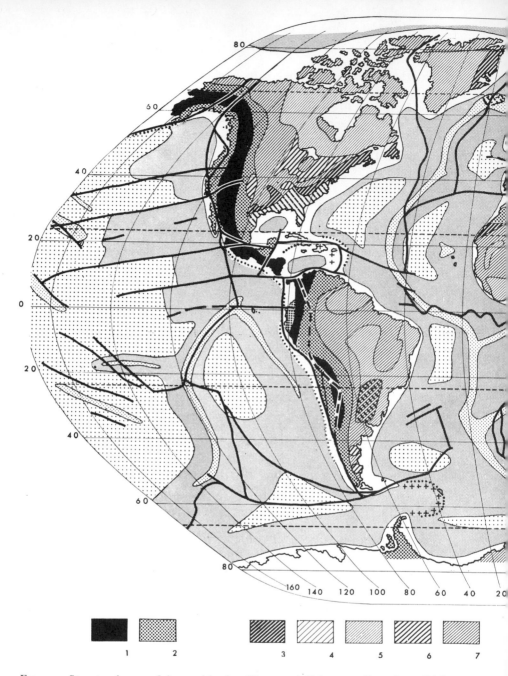

Fig. 1.4. Structural map of the world, after Glangeaud, Beloussov, Gutenberg, Richter and others

Alpine geosynclinals:
  1. Hercynian massifs caught up in the alpine geosynclinals
  2. Sedimentary zones of the alpine geosynclinals

Platforms:
  3. Caledonian folds and hercynian *antéclises*
  4. Hercynian *synéclises*
  5. Pre-cambrian shields
  6. *Antéclises* of folded pre-cambrian rocks
  7. Tablelands of folded pre-cambrian covered by Palaeozoic or Mesozoic strata

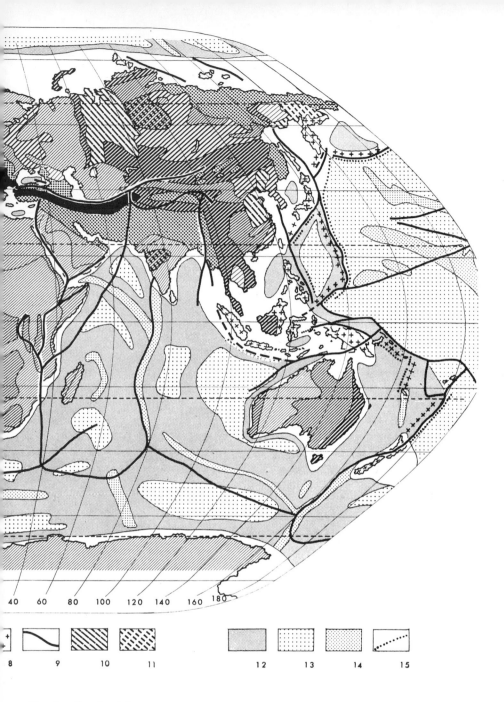

40   60   80   100   120   140   160   180

8   9   10   11   12   13   14   15

Margins of continental areas:
  8. Volcanic island-arcs
  9. Major fractures (margins of plates)
  10. Regions of major post-hercynian subsidence
  11. Principal areas of basalt flows

Oceanic Basins:
  12. Regions of medium depth, with intermediate or oceanic crust
  13. Abyssal plains
  14. Mid-oceanic basalt ridges with volcanoes
  15. Major ocean deeps

would be formed in Pre-Cambrian times by the consolidation of the upper crust into several nodes, which are the 'shields' around which the various continents have been formed. Indeed, the intensity of the Pre-Cambrian folding is such that if the folds were all flattened out the rocks would cover the entire globe. From this time, therefore, as Fourmarier has maintained, there have been permanent continents and ocean basins. A consideration of isostasy has led this author to formulate a law of the permanence of volumes; the uplift of one section of a continent must be balanced by depression elsewhere. Geological history provides us with many examples of such phenomena, which may not be exactly synchronous, but like isostatic reactions, may be staggered in time.

In each of the continents a rigid 'shield', resistant to subsequent folding, and formed of Pre-Cambrian rocks folded and hardened by metamorphism, constitutes the core against which successive waves of mountain-building, of uplift and depression, have broken. As Chevalier admits, centrifugal force would be capable of initiating some limited displacements, of which transverse faults are a geological proof, together with the tensions which gave rise to the mesogeic mountain chains and even perhaps the torsion observed on the various continents. But other forces could also cause folding. The edge of the continental areas is a major zone of discontinuity in the earth's crust. Several authors, notably Joly and Goguel, have pointed out that the sial is particularly rich in radioactive minerals, the disintegration of which would generate heat. Perrin and Roubault have regarded metamorphism as the motive force in folding. These ideas are linked, and some authors like Vening-Meinesz and Griggs have postulated convection currents on the edges of continental areas, capable of breaking up the edge and forming island arcs. Beneath the continents, these currents might even produce some erosion of the base of the sial. Gidon (1963), in a carefully moderated thesis, has formulated rather less spectacular proposals. Whatever may be the exact nature of the mechanisms involved, and this is still obscure, it is clear that the edge of the continental areas is a sensitive zone and that the difference in the nature of sial and sima develops forces at the zone of contact which are favourable to tectogenesis and especially to folding—hence the location of volcanic belts and island arcs. In the more stable areas, the continents terminate in a flexure (the 'continental flexure' of Bourcart), the orographical expression of which is the continental slope, more or less masked by sediments. Often the slope is carved by submarine canyons, and some of these have been dated and attributed to subaerial river erosion. There must therefore have been a depression of the continental edge as a result of the flexure.

Although still incomplete, our view of the main structural outlines of the earth is now reasonably clear. The balance between continents and ocean basins is a fact of prime importance. It implies a profound differentiation in structure which has been permanent, in general, since the Pre-Cambrian. The great mountain chains have been formed within the framework of this permanence, on the one hand through the contact of these differentiated parts of the earth's crust and probably under the influence of the forces to which this contact gives rise, and on the other hand, it seems, through the squeezing caused by centrifugal forces that tend to draw the continents towards the equator. Whatever their exact nature, all these phenomena owe their origin to the structure of the globe and to its internal dynamics. This is why the key to the gross physiography is provided by geophysics. At the same time the interaction with external forces is apparent. When erosion slowly lightens the continents, it initiates isostatic compensation. By modifying the equilibrium of the masses that constitute the earth's crust, it thus influences the geophysical mechanisms. As Cailleux has truly remarked, erosion must be compensated by continental uplift, otherwise the continents would be levelled off, in the long run, by gravity!

Certain very marked climatic changes, such as those connected with glaciation, may also cause relatively sharp and rapid isostatic movements. The Quaternary era probably witnessed five separate glaciations: recent studies lead increasingly to the conclusion that an earlier glaciation, the Danubian, antedated the four classical ones described by Penck and Brückner (namely, in chronological order, Günz, Mindel, Riss and Würm). It is probably in this direction that we must look for explanations of the present sea level, which is not quite what we might expect on purely structural grounds.

Variations in sea level and coastal changes

The actual position of the seashore at the present time constitutes something of an anomaly. It does not correspond to a single point on the hypsographic curve as one might have expected. The anomaly may be briefly stated thus; the oceans have invaded that part of the continental areas which corresponds to the continental shelf. This fact must be put into geological perspective in order to understand it and measure its geomorphological consequences.

The difference in density between water and air means that the distribution of land and sea interferes with isostasy, for a marine transgression increases the weight of the sial and thus tends to depress it into the sima. Conversely, a retreat of the sea will provoke isostatic uplift. This is why the general sea level is bound up with the structure of the earth's crust. This is not all, for sea level is also influenced by climatic variation. The accumulation of continental ice caps during the cold periods of the Pleistocene, and much earlier, in the upper Palaeozoic and lower Cambrian, could only take place at the expense of the volume of water contained in the oceans. And this phenomenon in its turn, involving a transfer of mass on the earth's crust, provoked isostatic adjustments. There is thus an interplay between sea level variations and isostasy. But the nature of these interactions is different from those hitherto considered, for external and not internal forces are their cause.

The mechanism of eustatism

The eustatic theory was propounded in the nineteenth century before the effects of glaciation on sea level had been appreciated. It was formulated especially in order to account for the great marine transgressions and retreats observable in geological history; and this reduces its interest for the geomorphologist. What are of most interest to us are the recent changes in sea level, and these are the domain of glacio-eustatism; so we shall restrict our discussion to the mechanisms, the character of which applies throughout geological time.

EUSTATIC EQUILIBRIUM

Eustatic equilibrium is closely bound up with isostatic equilibrium, through the medium of the weight of water on the ocean floors. This equilibrium is disturbed by tectonic movements. When a large folding movement takes place and causes a ridge or an island arc to rise from the sea, it reduces the area occupied by water. Conversely, if subsidence takes place, some land is drowned and the water area increases. Water, being extremely mobile, immediately adapts itself to the new situation. Modifications of the continental areas, however, provoke compensatory movements at the base of the sial, which are much slower to start and to cause displacements of matter. In other words, isostatic equilibrium is very slow of realisation, whereas eustatic adjustment is immediate.

The eustatic theory put forward this idea, which was elaborated later after the development of the notion of isostasy. During periods of tectonic paroxysm, the continents tend to rise and to bulge, and this causes a recession of the sea. This swelling is accompanied by deepseated movements which displace sima from the ocean basins towards the continents, where orogenesis results. The ocean basins are thus depressed. Fractures may occur that either lead to the formation of deep trenches, or enable magma to rise and form a chain of volcanic islands. The net result is a contraction in the size of the oceans, which sets in motion a recession; the greater mean depth of the oceans provokes a reduction in their area. Mountain-building periods are thus periods of oceanic retreat.

When it was first formulated, the eustatic theory was influenced by a certain conception of tectonic history that had been the basis of Davis's cycle of erosion. This was the notion of catastrophism, which viewed the earth's history as a series of rhythms—periods of intense tectonic activity separated by long periods of calm. This notion, essentially anthropomorphic and with roots that are to be found in the Bible, was put forward by Cuvier, elaborated by the German Stille, and used by Penck as the basis of his theory of the *Escaliers de Piémont* (*Piedmonttreppen*); but it has now been largely abandoned. The idea of a succession of scenes separated by curtain falls has been replaced by that of a more continuous evolution, though of variable intensity, with paroxysms separated by periods of lesser activity. It has also been realised that, contrary to the ideas of Stille, the tectonic paroxysms were regional, not universal, in their effects. Thus the massive orogenesis of late Jurassic age in the western United States has almost no equivalent in the Alps apart from small localised movements in the Briançon region.

The absence of global orogenic periods considerably hampers the scope of the eustatic theory. Given the considerable volume of the oceans and the comparatively small mass of the mountain ranges and of the subcrustal transfers of material accompanying the regional orographic paroxysms, it is difficult to see how the latter could have had much effect on the general ocean level. And besides, advances in our knowledge of world stratigraphy have destroyed the old notion of large-scale transgressions and retreats of the sea. For example, the great transgression of the middle and upper Cretaceous, formerly considered to be global, now appears to be limited to the Hercynian region of Europe. It did not touch the massifs of Brazil or West Africa, any more than the western coast of South America. It seems on the other hand, that the great orogenic paroxysms generated subcrustal currents that only affected the neighbouring plateau and plain areas, raising and lowering them alternately and so causing advances and retreats of the sea level that were limited to the part of the continent involved. In the Cretaceous of Hercynian France, for example, there is first a retreat, coinciding with folding in the Pyrenees and Provence, then, in the Albian, a strong and more general transgression while the paroxysm was at its height. Nothing comparable is found in the Alpine geosyncline. Rather than general variations in sea level in the geological past, there have been local tectonic movements both in time and in space, that had their effects on the position of land and sea.

The clearest examples of advances and retreats of the sea are to be found in the Mediterranean during the Neogene, with the Pontian retreat and the Plaisancian transgression that followed it. In many areas, there was time for deep valleys to be cut during the retreat, and the transgression converted these into rias: But this elegant example has no general significance, for it is purely regional. There is nothing like it in the Hercynian areas only a few hundred kilometres distant. The explanation lies in the peculiar character of the Mediterranean, which was a closed sea until the formation of the Strait of Gibraltar at the very end of the Tertiary. This relatively restricted mass of isolated water was apparently greatly influenced by tectonic movements in the area during the Neogene paroxysm, so that it is an exceptional case, of no general import. We can hardly, despite the work of Baulig, make much use of the eustatic theory in the pre-Quaternary periods. The explanation of the 'staircase' of erosion surfaces in any region does not lie in eustatic variations in sea level combined with the Davisian erosion cycle. Their doubtful existence depends on illusions derived from false graphical devices. As an example we may cite the 180 m and 280 m Pliocene levels which, in the Paris Basin, were cut by an imaginary sea that has left no deposits at all.

If eustatism is to be relegated to the rank of useless theory, it is by no means the same with glacio-eustatism.

GLACIO-EUSTATISM

Great changes were effected by the Quaternary glaciations, which were events of primary importance in the earth's history, the likes of which had not been seen since the Carbo-Permian. The glaciation resulted from the general, albeit unequal, refrigeration of the globe, the causes of which could only be cosmic. The forces set in motion were considerable, for the variations in the amount of heat received by the earth were enormous. And they were sufficiently powerful to upset the equilibrium of the crust.

The accumulation of enormous masses of ice, averaging 2 000 to 3 000 m thick, over some 20 per cent of the land surface, had two major consequences:

1. A general lowering of sea level occurred during each glaciation. It must be assumed that during a glaciation, which may have lasted for tens of thousands of years as in the case of the Würm (Weichsel), the total quantity of water available (whether frozen or liquid) on the earth's surface remained reasonably constant. By reason of the general lowering of the temperature, there must have been a corresponding lowering in humidity of the atmosphere and in its total water vapour content, but the bulk of the frozen water immobilised in the ice sheets must have come from the oceans. There is general agreement that in the Würm period this ice, if melted and returned to the oceans, would have raised their level by 100 metres. Each glaciation thus entailed a retreat of the sea (the pre-Flandrian recession in the case of the Würm), whilst each deglaciation resulted in a transgression (the Flandrian transgression at the end of the Würm in the late glacial period). Given the mobility of water, these phenomena would be produced simultaneously, without any timelag. In our own time indeed, the melting of the glaciers between 1900 and 1950 was accompanied by a rise of sea level amounting to 1·2 mm a year which is appreciable though much smaller than the maximum of the Flandrian transgression, which was about 5 to 6 mm a year.

2. The isostatic reactions set in motion by the important transfer of mass represented by the storage, in the form of ice sheets, of water derived from the oceans, and then by their return through melting to those same oceans. During the glaciation, there is a lightening of mass over the ocean basins and an increase beneath the ice caps. Conversely, with deglaciation there is a lightening of the glaciated lands and an increase over the ocean basins. The rising of areas formerly glaciated has been observed and measured

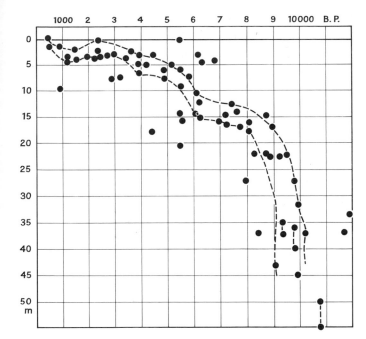

FIG. 1.5. The Flandrian transgression, based on C$_{14}$ dating
(*from H. Graul, 1959*)

On the horizontal axis, time in thousands of years B.P. (before present). On the vertical axis, altitude of marine beaches above present sea level. Each dot represents one measurement. Points away from the main cluster may indicate either errors of measurement or local movements (of tectonic origin, or due to the settling of unconsolidated sediments).

A remarkable rise in sea level from 10 000 B.P. to 8 000 B.P. will be noticed; this corresponds to the massive melting of the last relics of the Scandinavian ice-cap and more particularly, and at a slightly later period, of the North American ice-cap. Then, from 8 000 to 6 000 B.P., the rate of rise slackened, owing to the isostatic deepening of the oceanic basins resulting from the weight of water entering them from the melting ice. The transgression resumed during the thermal optimum between 6 000 and 4 000 B.P., that caused an increased melting of the remaining ice-caps and glaciers.

and constitutes one of the principal verifications of the theory of isostasy. It is therefore natural to suppose that similar modifications must occur in the ocean basins, although we cannot directly observe and measure them. In both cases there is a timelag of a few thousand years. As deglaciation proceeds, the meltwater returning to the oceans causes a rise of sea level. But then isostatic recovery starts, and this tends to depress the ocean beds and raise the level of the deglaciated areas. The meltwater from the glaciers represents but a small depth when spread out over all the oceans (about

53

100 m in the Würm, and remembering also that the water has a density only one-third that of sima), whilst the weight of the ice, concentrated over very much smaller areas, was considerable (20 or 30 times greater per square kilometre). The isostatic adjustment of the ocean beds could thus not have the same amplitude as that of the formerly glaciated areas. During the early part of the deglaciation, the threshold is not reached and the ocean beds are not lowered; all the water liberated by the icemelt is available to raise the general sea level, and transgression, unhindered by any counterreaction, is rapid. But when the sea level has risen by some 70 to 80 m, the threshold is reached and isostatic reaction begins. The ocean beds are depressed by the weight of water and the deepening of the sea acts in opposition to the transgression, which thus slows down. This has been documented in detail for the Flandrian transgression of which the various phases have been dated precisely by C_{14} methods and correlated with the melting of the ice sheets.

There is thus a profound difference between eustatism and glacio-eustatism. Eustatism pure and simple is an improbable theory unsustained by any serious argument and it is very doubtful if it could have the geomorphological implications that have been ascribed to it. On the contrary, glacio-eustatism, like isostasy, is supported by direct observation, as of recent sea levels correlated with glacial retreat, or by calculated inference, as with the precise dating of the Flandrian transgression. It must thus be considered as basic to geomorphology. But it must not be treated in isolation, for it does not act alone. It reacts with isostasy once a certain threshold is reached, and this has happened with all the major glaciations. The combination of the two phenomena provides the only explanation of certain established facts, such as the way in which, in glaciated areas, an early transgression is followed by a retreat of the sea. The early transgression in Sweden, for example, that accompanied the deglaciation and for a large part of this period caused the glacial margin to be occupied by the sea, happened at a time when the general sea level had already risen perceptibly as a result of the deglaciation, but before isostatic adjustment had begun. When the latter did at last occur the land rose more rapidly than the sea, the more so since after a certain time there was isostatic depression of the ocean floors. Thus the transgression was followed by a retreat, which is still in progress. Such phenomena, in essence geophysical, are at the root of the evolution of landforms.

The combined influence of isostasy and glacio-eustatism was operative during the whole of the Quaternary, which begins with the first glaciation. The entire period is thus characterised by repeated changes in the relative level of land and sea and so of oscillations in the general base level of erosion that have influenced the evolution of landforms. The geomorphologist must of necessity interest himself in these changes, and to this topic we now turn.

Oscillations of sea level during the Quaternary

One of the characteristic features of the Quaternary is these rhythmic oscillations of sea level, corresponding to climatic variations. The sea level changes have influenced morphogenesis through frequent changes in the base level of erosion, while the climatic changes have resulted directly in alterations in the nature of the erosion processes. Morphogenetically therefore, the Quaternary is an exceptionally complex period, very different, for example, from the Tertiary, in which changes of sea level were local and brought about by tectonic deformations. Such earth movements did not cease in the Quaternary; far from it. The isostatic reactions to the transfer of mass during the glacial periods had the effect of increasing the mobility of the crust by creating frequent disequilibria which were added to the general play of tectonic forces. The repercussions of climatic changes and oscillations in the general base level are thus just one contribution to the morphogenetic peculiarities of the Quaternary, the fundamental cause of which was a variation in the amount of solar energy received by the earth.

The complexity of these interactions demands precise methods of study, and we may briefly outline some of these before expounding the general conclusions.

CORRELATION OF FORMER SEA LEVELS

This is a field of study in which many false methods have been used, leading to outmoded conclusions.

In any given region, changes of sea level result from two sets of movements: general oscillations of sea level and local tectonic deformations. According to the circumstances, the two types of movement either reinforce each other or are in partial opposition. In general, changes of sea level have been rapid and widespread during the recent past, with the Flandrian transgression taking place between 12 000 and 2 000 BP. They have thus had every opportunity of masking the regional or local effects of tectonic movements, and this is what has generally happened. Since tectonic forces have had little time to operate, it is rare to find the levels of the Flandrian transgression deformed, and they are found at the same altitude almost the world over. However, in areas of active subsidence, such as the delta of the Mississippi, they are normally at a lower level. On the other hand, as one goes progressively further back in time to examine older sea levels, differences in altitude between one part of the globe and another increase; in the older Quaternary, divergences are the rule. Tectonic movements, which are generally slower than the Flandrian transgression, take on a progressively more important role with an increase in the time they have had in which to operate. But tectonic movements themselves vary, of course, from region to region. We thus find old beachlines at 100 m above present sea level at various localities round the Mediterranean, and these have been regarded as remnants of a general level. But if the sea had in fact been at this level, it would have invaded vast areas of the Aquitaine basin, the plain of Languedoc, the lower Rhone and Senegal, areas in which detailed studies have revealed nothing of the kind.

The situation is thus really very simple: the older the sea levels the more likely it is that they may have been deformed. Recent sea levels are concordant on all coasts except those in which tectonic activity is still intense, such as Japan and New Zealand: but the levels of the lower Quaternary have been generally modified by tectonic movements and are therefore very varied, indeed, quite divergent. To say that they are not distorted leads to absurdities of the same type as the application of eustatism to the Massif Central or the Paris Basin in the Tertiary epoch.

It is often difficult, however, to determine whether or no a sea level has been affected by tectonic movement. Beaches and abrasion surfaces have always an original slope, which is a function of wave action and the nature of the rocks, and it is difficult to determine their theoretical value. In the same way, it is difficult to say precisely whether an old marine erosion surface is deformed or not. Subsequent dissection, which may have obliterated the original landforms, may increase the uncertainty. Furthermore, as we have already noted, sea levels or river terraces that have been affected by tectonic movements, as in Venezuela, Chile and Peru, have been raised or lowered vertically over relatively small areas by folds and faults. Within one such area the levels do not appear to be disturbed. They may be slightly warped, but their slope differs too little from an original slope for one to be able to determine whether or not it really is one. But only a few kilometres further on, the terraces may slope at 20° or 30°; and along folds and faults erosion has been especially easy, so that the material has been spread out from its original narrow band, making identification difficult.

Under such conditions it is easy to see why many authors have been misled and have regarded systems of terraces as undeformed whereas in fact they have been subjected to important vertical movements, with warping insufficient to be readily distinguished. This is often the case in the older Quaternary. A critical mind, precise measurements, and a constant regard for other data and other regions are necessary for understanding the correlation of old sea levels. These things are often absent. At the beginning of this century, the Davisian concept dominated studies. Thus Deperet and de Lamotte assumed an interrupted eustatic lowering of sea level throughout the Quaternary. Starting with a succession of levels established without much care in the vicinity of Algiers and extended into Tunisia, Sicily and Calabria, they postulated the universal existence of marine levels at 180 m (Calabrian), 90–100 m (Sicilian), 55–60 m (Monastirian), 30–35 m (Tyrrhenian), 15–18 m and 6–8 m. Baulig subsequently modified this interpretation and showed, after a study of the Crau region, that the sea level did not lower gradually in between these standstills, but on the contrary was lowered much more, to rise again later. These intermediate retreats were attributed to the glaciations, and glacio-eustatism was born. The demonstration rested on the behaviour of the deltaic deposits of the Durance, in the Crau region, which are of fluvioglacial origin but plunge beneath the Camargue and beneath the sea. Baulig attributed it to the Würm, but in fact it is Mindel or even earlier.

In order to make a valid reconstruction of old sea levels, and of river terraces, we must have recourse to a combination of methods that excludes the risk of entering a vicious circle. Mere altitudes cannot provide a basis for dating. We must proceed in the following manner.

The first task is to establish exactly the succession of the various stages in the geomorphological evolution, to define the relations between the various levels and to investigate the phenomena that were produced during the intervening periods, especially the erosional phases. It is generally necessary not to limit one's attention to the marine flats, but to study also the neighbouring strands and slopes and the streams that cross them, in order to correlate the beachlines and river terraces. It so happens that during recessions, the sea usually retreated some distance, often far beyond the present shoreline, so that the former beaches have suffered subaerial erosion, and cliffs have been transformed into slopes. A detailed geomorphological map is in most cases the surest way of appreciating such relationships.

The study of deposits, both in the field and in the laboratory, is an indispensable part of the investigation. It permits the identification of the various marine and terrestrial formations, and enables one to see to what degree one deposit contains material derived from another. In such cases it is clear that the deposit containing the derived material is younger than that from which the material came.

An examination of the palaeosols developed on the surfaces of the various geomorphological facets of the landscape also contributes to correlations and dating. Some such palaeosols have been dated over a wide region, and this enables individual observations to be fitted into a general chronology. Thus in the south of France, red earths *in situ* are at least as old as the Mindel–Riss interglacial; they are never found on the deposits of the Riss glaciation or more recent formations.

Once the succession has been established, one can proceed to relate it to the general evolution of the landscape. The possibilities vary greatly from one region to another, and it is always desirable, as far as possible, to make crosschecks between regions.

Radio-carbon analysis now permits absolute dating—though not, it is true, completely free from error—within the last 50 000 years. It gives us a general chronology of the Flandrian transgression and enables detailed correlation with the Würmian deglaciation to be made.

Palaeontology is of rather doubtful value, since Quaternary time has been too short to permit of much in the way of perceptible evolution amongst the littoral faunas. Species show more relation to change of temperature of the local sea water than to the passage of time. The study of the relation between the two oxygen isotopes O_{18} and O_{16} makes possible precise estimations of these temperatures. C. Emiliani was thus able to show that on the borders of the western Mediterranean, all the raised beaches—save for a few exceptions caused by tectonic deformations—contain molluscan faunas that lived in warm or tepid water. On the contrary, dredgings from submerged beachlines down to − 100 m in the Golfe du Lion show cold-water faunas.

Archaeological discoveries furnish further evidence, for human artefacts evolved more rapidly than molluscan faunas during the Quaternary. Some of them have been precisely dated, like the old cliff-caves that were subsequently occupied by man.

Finally, palaeosols and the products of ancient weathering also often have a regional significance and can be dated by archaeological means or by C_{14}. But in these cases one must take account also of the influence of location, of local climate and of the parent rocks.

It is by the application of all these methods that it has been possible to correlate raised beaches, to determine their relationships to palaeo-climatic changes through the study of terraces and slope deposits, and to estimate the part played by the tectonic warping in their present altitudinal position.

QUATERNARY VARIATIONS IN THE GENERAL SEA LEVEL

Variations in sea level during the Quaternary were controlled, if we accept glacio-eustatism, by glacial fluctuations. It is necessary also, however, to take into account isostatic reactions and changes in the volume of ocean water due to expansion with a rise of temperature and reduction when the temperature falls—changes which work in the same sense as glacio-eustatism. If we know, through palaeogeographical reconstruction, the extent of the ice sheets during the different Pleistocene glaciations, it should be possible to calculate the approximate volume of water locked up therein, and as a consequence the degree of lowering of the sea level. Naturally the calculations will be more precise for the most recent periods; they are probably quite accurate for the Würm and Riss glaciations, but are more doubtful for the older Quaternary—the Günz and Danubian.

At the present time although we are in an interglacial period, vast masses of ice remain on the earth's surface. Nine-tenths of the ice is in the Antarctic. But although it has been established that notable fluctuations of the Antarctic ice cap occurred during the Quaternary, it is not these fluctuations that played the decisive role during the glacio-eustatic recessions. Virtually the whole continent is ice covered, so that there was little opportunity for the advance of tongues of glacial ice, and the ice cap itself increased in thickness and became more domelike. This swelling was limited, however, by the increased outflow which resulted from it. The glacio-eustatic recessions of the sea during the Quaternary were due in the main to the forma-

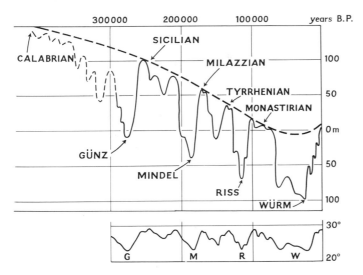

Fig. 1.6. Glacio-eustatic oscillations of sea level during the Quaternary (*after Fairbridge, 1961, p. 131*)

The graph is schematic, for absolute dating is only possible for the last 50 000 years (by C_{14} methods). The time-scale is perhaps too short, for other authors place the Calabrian–Villafranchian at between 2 000 000 and 700 000 B.P.; but on the oscillations there is a large measure of agreement.

A synchronism may be noted between (*a*) variations in sea level (upper part of graph) and (*b*) fluctuations in the temperature of tropical seas (lower part of graph).

The cold periods are thus universal phenomena of which the glaciations are but one manifestation. The variations in sea level show, in addition to the oscillations due to the cold periods, a slow general fall throughout the Quaternary (indicated by the broken line); this would seem to be due to the piling-up of ice in Antarctica, and perhaps also to general deformations in the continental areas and oceanic basins resulting from the glaciations

tion of new ice caps which have since disappeared, over Scandinavia, North America, and to a much smaller extent northern Siberia (especially the northern Urals).

During the Quaternary glaciations, the level of the sea was much lower than at the present time. But during the long and warmer interglacial periods, there was actually much less ice than remains at present, so sea levels were higher. The Greenland ice cap is but a relic, a survival, for it is not still growing. Indeed, it is in thermal disequilibrium with the present climate, for the ice is colder than the atmosphere, and if the present conditions were to last for long enough it would disappear. In contrast, the Antarctic ice cap, which is roughly in equilibrium with the climate, would persist. But the Mindel–Riss interglacial was very prolonged (probably 200 000 years), and probably warmer than the present. It is possible that the Greenland ice-cap would have disappeared during that time, whereas the Antarctic ice remained, though perhaps reduced in volume. At the beginning of the Quaternary, when the climate began to get colder, there were probably no ice caps. They grew gradually, lagging considerably behind the climatic change that caused their growth. An ice cap the size of the Antarctic would probably take some 50 000 to 70 000 years to form. It is possible that it was not in existence at the time of the Danubian glaciation, but was preceded by mountain glaciers which would form more rapidly and would respond more quickly to climatic changes. The present Antarctic ice-cap welds together two parts of the continent that are separated by a sea gulf. The filling of this gulf by ice—which is now 3 000 m thick—must have taken a very long time; for the present rate of ice accumulation is less than 15 cm a year and it was probably slower at first, when the altitudes were less.

Our knowledge of present and Pleistocene glaciers enable us to cross-check the observed facts relating to old sea levels, with the following results:

1. In the older Quaternary, sea level was higher than it is today. As we have seen, these old sea levels are the most difficult to reconstruct, as much by reason of their subsequent dissection as because of tectonic displacement. However, we can give them an approximate upper limit by calculating the extent of the rise of sea level that would take place if all the present ice were to disappear, because this is the situation that must have prevailed at the beginning of the Quaternary, before the formation of the Antarctic ice cap.

This involves the calculation of the masses of the Antarctic and Greenland ice caps; and the multiplication of geophysical surveys in recent years has enabled us to do this with some precision. We must, of course, also take account of isostatic recovery and of the expansion of the ocean water with a rise of temperature, but these are relatively unimportant factors. Estimates vary from about 30 to 60 m. Several authors who have studied ancient sea levels in areas which appear to have great stability agree that the maximum elevations of Quaternary beaches that have not been tectonically warped are of this order: 50 m in eastern USA (Wolstedt, 1960), 29–30 m in the Balearic Islands (Butzer and Cuerda, 1960), 12–30 m on an average (Cotton, 1963), 30–35 m in Madagascar (Battistini, 1962). The agreement is general on a maximum altitude of between 30 and 40 m. Levels higher than this must have been tectonically raised, notably in the case of the rather unhappily chosen Mediterranean examples quoted by Lamotte and Deperet, and in areas of vulcanicity and active earth movements. It is possible that the considerable disequilibrium caused by the lowering of sea level at the time of the formation of the first ice caps set in motion particularly violent isostatic reactions that would have upset the old platforms; this might indeed be the explanation of the widespread tectonic paroxysm that occurred in the early Quaternary, and of the frequency with which shorelines of that period have been deformed.

It is probable that the Danubian glaciation, and perhaps even the Günz, only caused a minor recession of the sea. It seems unlikely, indeed, that during the Danube–Günz interglacial the sea level would have been as low as it is at present. The first great recession below the present sea level occurred with the Mindel glaciation, for it was only at this time that vast ice caps were formed in Europe and North America. The Günz glaciation was of smaller extent, under an apparently less cold climate.

2. For the Middle Quaternary, data are easier to come by and the available records are numerous and more reliable. The Mindel–Riss interglacial (sometimes called the 'great interglacial'; or Hoxnian, by English authors) was of long duration and quite warm. It produced the last subtropical weathering in the Po Basin (the *ferretto* or red earth) and in Mediterranean France. The Antarctic ice cap probably persisted but in a much reduced form. So the sea level rose to about 18–20 m above the present level. This is the *Tyrrhenian I*, which corresponds in northern Germany to the *Holstein Sea*, the deposits of which are intercalated between the moraines of the Elster glaciation (= Mindel) and those of the Saale (= Riss). The sea level remained at these altitudes for a long time, with minor fluctuations. Then it fell gradually to just below its present level, at the onset of the Riss glaciation. A succession of pebble beaches and abrasion surfaces marks out this recession in some areas, as Guilcher has demonstrated in Brittany, for example.

3. In the upper Quaternary, the Riss–Würm interglacial was cooler than the preceding one, and also shorter—only some 50 000 years as against 200 000. The sea level rose, but did not reach the Tyrrhenian I level. It remained at 6–8 m above the present level. This is the *Ouljian*, as defined in Morocco by Gigout. It corresponds to the *Tyrrhenian II* of some authors and to the *Eemian* or *Ipswichian* of the North Sea borderlands. It would seem that during this period the Greenland ice cap had time to disappear almost completely whereas the Antarctic cap was simply thinned. In middle latitudes the temperature was probably not much higher than at the present time.

At the end of the Würm, the Flandrian transgression reached a level just slightly above the present, about 1·5 m. This is the *Dunkerquian*, which dates from the beginning of our epoch and has been of particular importance in the morphological development of deltaic and fluviomarine plains, especially in tropical areas, and of atolls.

4. The marine recessions set in motion by the various glaciations, from the Mindel onwards, are less well known, because of the obvious difficulty of studying submarine geomorphology. But here also, a correlation with the masses of ice locked up on the continents is possible.

For the Würm, the two methods give closely comparable results. It is agreed that the recession in stable areas was between 80 and 100 metres.

In a variety of regions, dredging has yielded beach material, notably pebble banks with shells, at such depths. Peat, formed on land, has been found at slightly shallower depth, as in the North Sea.

The Würm glaciation, however, was not the most extensive. In almost all parts of the globe the limits of the Riss glaciation exceed those of the Würm, and the corresponding recession was therefore presumably greater. But it is difficult to distinguish, among the submarine surface forms that were produced by subaerial erosion, those that pertain to the Riss and those that belong to the Mindel, which were also the result of a recession greater than the pre-Flandrian. So, very prudently, the expression 'maximum recession' is used. Kuenen (1954) puts it at − 120 m. Donn, Farrand and Ewing (1962) give a maximum depth of 150 m for the ancient submerged beachlines. Cailleux (1954) suggested figures of the same order. These depths correspond roughly to the level of the continental shelf, which over much of its extent has been fashioned by subaerial erosion, and which thus constitutes a very special part of the earth's crust, as suggested by the hypsographic curve.

Starting with purely geophysical data to account for the form of the continental areas and ocean basins, we have been led, as we delved into greater detail, to a consideration of climatic phenomena. Thus there is no doubt about the interaction of internal and external forces. Quantitative changes in the external forces, as in palaeoclimates, lead to qualitative modifications, in the form of glacio-eustatism, and these react on the internal dynamism through isostasy. But the impact of the external forces on the internal dynamism remains a subordinate phenomenon; its role is abnormally great in the Quaternary, because of the amplitude of the climatic fluctuations that characterise this period.

2
Geosynclines and fold belts

From the survey of the general surface form of the globe to which we now turn, it is evident that it is made up of an assemblage of three major structural units (macrostructures):

1. the bed of the ocean basins, relatively rigid and stable, but affected by tear faults where the sima is near the surface;
2. the continental masses, likewise rigid and stable, but subject also to some fracturing; made up of the sial, which may or may not have a sedimentary cover. These continental masses go under the name of *platforms*;
3. particularly unstable regions, where the majority of earthquakes occur, occupied either by mountain chains of the highest altitude, or by ocean trenches: these constitute *geosynclines*.

Geosynclines represent sensitive zones within the crust which are subject to particularly intense and violent tectonic forces. It is here that relief is most marked, gradients most severe and, as a result, the gravitationally induced forces of dissection are the most active. Compressive forces produced in the geosyncline result in folded rocks. Usually, fold belts originate in geosynclines. All the great fold mountain belts of the earth are the product of geosynclinal evolution. Nevertheless, the relationship is not an absolute one. Folding may also affect the sedimentary cover of the continental platforms on the margin of the geosyncline. While the Alps are a mountain chain of geosynclinal type, the Pre-Alps and the Jura are made up of folded continental platform structures, i.e. of folds in the sedimentary cover (*plis de couverture*).

In geosynclines, the intensity of the deformations has given rise to marked tectonic relief. The geomorphological features faithfully express the tectonic features. Moreover, the rocks often show little variety. The vigour of the orogenesis, the upheaval which gives rise to mountains, is matched by the severity of the dissection. The direct influence of tectonics is of first importance and it is here that geomorphological study must begin. In regions of folded cover rocks, however, strata are generally more varied lithologically, and their deformation less severe. Here also, tectonics determine the landforms which develop, but in a manner which is less strict and exclusive. As in the continental platform regions, movements of the whole rock assemblage often set in motion differential denudation. The characteristic forms of Jura type relief are thus produced. There is reason, therefore, to distinguish between geosynclines and folded cover rocks, and also between the tectonic and the geomorphological approaches.

Accordingly, we shall study the evolution of geosynclines, the relief forms of geosynclinal fold belts and, finally, the characteristics of folded cover rocks.

Evolution of geosynclines

The geosynclinal concept was elaborated almost a century ago by Hall and Dana, then by Haug, in order to account for the origin of the great mountain chains. From the beginning, these authors were struck by the fact that mountain chains are formed in regions where accumulated sediments are particularly thick. This is why, under the influence of the catastrophist ideas then favoured, they envisaged an initial phase of downwarping or subsidence, favourable to the accumulation of great thicknesses of deposits, during which folding developed at depth. This is the phase of *tectogenesis*. Then, in the second phase, the geosynclinal region was uplifted and rose in the form of a mountain chain, *the orogenic phase*.

Since that time, a great number of studies have been made and the accumulation of observations has necessitated some modification of this scheme. A general theory of geosynclinal evolution, with variants, has been established, mainly by the Russian authors Beloussov, Peyve and Sinitzyne. This will be adopted here, together with sundry results of workers in other countries. The theory distinguishes three periods named after the predominant characteristic of each: downwarping; formation of mountain cordilleras; uplift and stabilisation. We shall discuss them in turn and then examine several alternative views which have appeared from time to time.

The period of downwarping and heavy sedimentation

In certain parts of the globe shearing of the crust gives rise to zones of collapse. They result essentially from dragging forces and assume an elongate form, usually curved. From these qualities arise the two essential characteristics of fold belts: their elongation and their curvature, the latter often resulting in their disposition in festoons or arcs. The breadth of these shear belts is of the order of about one-tenth of their length.

ORIGIN AND DISTRIBUTION OF GEOSYNCLINAL TROUGHS

Geosynclinal troughs occur in two different kinds of locality, which seem to correspond to two distinct origins. As with the great mountain chains, some are found on continental margins while others lie between two continental blocks. The former are asymmetrical or unilateral (*liminaire*) troughs, and the latter intracontinental, symmetrical or bilateral (*biliminaire*) troughs.

The unilateral troughs

These are particularly well developed all around the Pacific and give rise to a series of ocean trenches which mark the borders of the continental blocks or run parallel to the margins of the island arcs. While there are many explanatory hypotheses, it appears that tension resulting from contact between sial and sima is the cause, as one indication of this is the strong seismic activity which characterises these zones. The general tendency for the continental masses to be uplifted, allowing them to remain high relative to the oceanic areas, demands compensation by subcrustal currents. These conditions may account for the origin of shearing or downdragging effects, mainly on the continental margins.

The intracontinental troughs

The continents themselves are sometimes subject to considerable tensional stresses resulting in dislocation, as is the case, for example, in the East African rift system. These shears initiate the development of geosynclinal troughs. They are found mainly in the Old World, extending from Gibraltar to Assam (the mesogeic belt). In contrast to unilateral troughs, they are bordered on both sides by continental forelands, whence is derived the term bilateral. In some cases, notably in the mesogeic zone, it appears that tensional stress due to the centrifugal force of the earth's rotation must be invoked to explain them. Nevertheless, the unilateral troughs coincide essentially with the contact between the Pacific basin and the neighbouring continental blocks, and the intracontinental troughs with the mesogeic zone of the Old World. Those geosynclinal troughs which are actively developing at present lie close to those which were active in the geological past and gave rise to mountain systems in Mesozoic and Cenozoic time. This is an expression of the continuing structural evolution of the earth.

However, not all such troughs do give rise to geosynclines. For this to be so, they must be the centre of intense sedimentation.

SEDIMENTATION AND THE APPEARANCE OF GEOSYNCLINES

Only those downwarps which receive abundant sedimentation are transformed into true geosynclines, giving rise thereafter to mountain chains. Moreover, only detrital materials can bulk large in such a sedimentation, and it is in this situation that downwarping has a major role to play.

Undeveloped geosynclines

Some downwarps do not give rise to true geosynclines, simply for the want of sedimentation. These are termed undeveloped geosynclines. Examples are provided by the deeper ocean trenches of the present era. Where these occur, the earth's crust is strongly downwarped. But, due to their position (in relation to the continents) these great depths receive almost no sediments. They lie too far from the coastlands for material of terrestrial origin to reach them. Sedimentation derived from living forms is also very slight, debris of crustaceans dissolving slowly in sea water in the course of the long transit into the ocean depths. The oceanic red clay which carpets the great depths is made up solely of insoluble residues and accumulates extremely slowly. When these ocean trenches lie adjacent to arcs of small islands, the situation is little different, for the detrital material furnished by them is of very small volume. In these conditions, downwarping yields a submarine trench which persists only as long as the tectonism which give it birth is maintained, provided that, in all cases, tension does not result in rifting. In such an event, magma may be released and the trench become filled with lavas which may build up into an arc of volcanic islands. Furthermore, a new trench, whose formation is probably favoured by the subsidence related to the emission of lavas, may develop at the foot of this island arc.

Such an evolution in undeveloped geosynclines, resulting in the development of submarine trenches and volcanic arcs, is characteristic of areas of subsidence remote from sources of detrital sediments, i.e. the coasts.

Sedimentation in geosynclines

The intracontinental trenches or the unilateral trenches lying close to the land masses, on the other hand, are favourably placed to receive abundant sediments, facilitating the evolution of true geosynclines. Moreover, the emplacement of these sediments varies according to local circumstances.

Submarine slumps and turbidity currents, made up of the flow, along the ocean floor, of water turbid with sediment and therefore of greater density than the clear water beneath which it flows, are of major importance in some ocean areas. For example, it has been shown in the Gulf of Mexico that the mass movements which occur periodically on the frontal slope of the Mississippi delta, notably during earthquakes, result in turbidity currents which transport some debris for distances of 500–1 000 km and into ocean depths of 2 000–3 000 m. Similar phenomena occur in front of the majority of deltas. They produce distinctive sediments, termed *turbidites*, made up of fine sands, silts and clays which are deposited by natural sorting in beds ranging in thickness from a few millimetres to several decametres to produce graded bedding. The coarser grains are deposited first, progressively finer grains being deposited where the turbidity current spreads out and terminates. This type of sedimentation is extremely common in geosynclines. It is found notably in some Alpine flysch deposits and in the molasse sediments of the Alpes Maritimes. It may accumulate in thicknesses of a thousand metres or more. It is not uncommon for turbidity currents to transport plant debris or littoral crustaceans, even freshwater

66

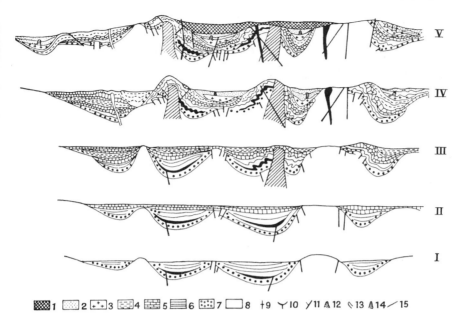

FIG. 2.1. Schematic outline of geosynclinal development according to Beloussov (1951)

1. Rocks accumulated in the furrows occupying the sites of consolidated geosynclines within the platforms
2. Molasse deposits of piedmonts or intermontane basins
3. Saline or gypseous deposits
4. Piedmont deposits, coal- or oil-bearing
5. Limestones
6. Sandstones and slates
7. Coarse sandstones and conglomerates.
8. Pre-geosynclinal continental basement
9. Terrestrial volcanoes
10. Ophiolitic intrusions
11. Mineralised veins
12. Granite batholiths
13. Small alkaline intrusions
14. Salt cones (diapirs)
15. Faults

The geosyncline is initiated on the platform by the formation of furrows which are of the bilateral type in this case. Earth movements, mainly of vertical type, give rise to compressive stresses as the wedge-shaped masses readjust; this, in turn, produces folding

types, into submarine trenches, a situation which remained an enigma for a very long time.

At other times, principally in the case of intracontinental rifts, sedimentation is continental rather than marine. Alluvial plains, occupied at one time or another by lakes, develop on the site of subsiding trenches. Conglomerates and gritstones accumulate here, sometimes with intercalations of very fine-grained material, and even evaporites may occur if the climate is arid. Debris cones and fans of finer debris overlap one another. This type of sedimentation has prevailed, for example, in some central Asiatic geosynclines since the end of the Palaeozoic. Generally, sediment is more easily supplied along the margins of the rifts, especially as the development of cavernous hollows causes blocks to tilt between the faults, which often leads to the production of relief, one of the ends of the tilt block being elevated by pivoting.

The geosynclinal trough does not necessarily give rise to a submarine depression, therefore. If sediments are sufficiently abundant, sedimentation can keep pace with the downwarping and maintain a fluviomarine plain, emerged or gently inclined beneath a shallow sea, like the Po delta in the Adriatic Sea. This is the case particularly where sharp relief borders the downwarp, a situation which is more frequent during the cordillera phase than during the initial phase here discussed. The debouching of a large, sediment-rich river, such as the Mississippi, into the geosynclinal trough, favours an accumulation of turbidites. In contrast, the troughs which develop too far from the coasts result only in undeveloped geosynclines.

The Cordillera phase

The initial phase of downwarping lasts only from 10 to 20 or 30 million years. It demands, in effect, a permanent crustal tension in the region, tension which permits gradual downwarping. This state of things implies a lack of isostatic equilibrium, in the form of a negative gravity anomaly: the volume of sea water and fresh, little-compacted detrital sediments which occupy the geosynclinal trench have a density which is much lower than that of the sial and very much lower than that of the sima. The downwarp itself, therefore, engenders an opposed force, isostatic recovery, which tends to check the downwarping and even to promote uplift. The second phase of geosynclinal development is characterised by this conflict between regional tension tending to maintain the downwarping, and isostatic recovery producing uplift to compensate for the deficiency of mass.

THE OPPOSING ROLES OF REGIONAL TENSION AND ISOSTASY

In the beginning, as we have seen, regional tension gives rise to a geosyncline in initiating major downwarping along a line of weakness, a rift, which is typically arcuate in plan. As long as the downwarping is slight, the isostatic disequilibrium is slight also. On the other hand, when it reaches many thousands of metres in depth, the isostatic disequilibrium becomes considerable.

Nevertheless, this isostatic disequilibrium varies a good deal from one geosyncline to another and from one part of a geosynclinal trough to another. In essence, it is a function of two factors:

1. The downwarping of the geosynclinal trough with respect to neighbouring portions of the crust. A submarine trench falling to −9 000 metres, amid abyssal plains in isostatic equilibrium around −5 000 metres gives rise to an isostatic imbalance of the same order as 12 000 metres of sediments in an intracontinental trench whose borders are made up of sialic material.

2. The sedimentary fill of the geosynclinal trench. This is never equally distributed : areas of rapid accumulation, such as the deltas, alternate with areas of slow filling (distal areas). When the coarse sediments (sands and gravels) are well sorted, they do not become compacted to the same extent as the fine sediments (clays and silts). Accordingly, given uniform downwarping, isostatic imbalance is greater in regions where accumulation is slight and predominantly of fine material.

On one hand, therefore, we have a difference in intensity between the tendency to downwarping, due to regional tension, and isostatic recovery tending toward uplift with, on the other hand, the differences in the amount of isostatic imbalance from one part of the trench to another, controlled by irregular sedimentation.

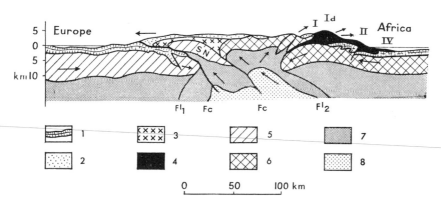

FIG. 2.2. Schematic diagram of a bilateral tectogene : the Spanish–Moroccan assemblage (*after Glangeaud, 1957*)
1. Autochthonous material of the Rif and Sub-Betic furrow
2. Malaga nappes
3. Median *mélange* of Las Alpujarras
4. Nappes of the Rif
5. Cratonic sial of the European block (platform of the Iberian Meseta)
6. Cratonic sial of the African block (eastern Moroccan platform)
7. Lower part of the crust and upper part of the mantle
8. Counterbalancing sialic mass
The diagram shows the tectogene phase with the formation of mountains on the site of the geosynclinal furrow. Rupture of the crust and rise of the sial occur at depth. At the surface, compression develops between the two rigid blocks of the European (left) and African (right) platforms which gives rise to the uplift and outward thrusting of the nappes

As tensional stresses over a region are not acting on uniform material, they are not uniformly distributed in the geosynclinal trough, any more than the tensions resulting from isostatic imbalance. Regional tension is expressed for the most part in a plane tangential to the earth's surface, while tensions due to isostatic factors act essentially normal to the surface of the earth. From the moment these opposed forces develop, the geosynclinal trough is subjected to considerable stresses which tend to break it up into sections. Ruptures are produced which are often strongly influenced by old lines of weakness. They delimit sections of the geosyncline which act differentially one to another. Some sections resist downwarping or even become uplifted. In marine geosynclines, the folds thus produced sometimes emerge above sea level. In terrestrial geosynclines, they result in parallel ridges, forming belts of small mountains. These are the *cordilleras*. Opposed to this are some other parts of the geosyncline which continue to subside: these are the geosynclinal furrows (*sillons*) smaller then geosynclinal troughs and the result of differentiation of the geosyncline during this second phase of its development.

Of course, the time required to reach this stage varies considerably from one geosyncline to another. It is a function of the combination of three factors:

1. The intensity of tangential tension arising from regional tectonism. This differs in the circum-Pacific belt from that of the mesogeic group because the origin of the tensions is different in each case.
2. The resistance of the underlying structure to the stresses to which it is subjected: some structures break up more readily than others, some allowing the formation of deep fractures from whence come magmatic outpourings while others fail to do so, and so on.
3. The isostatic imbalance which is a function of the nature and distribution of the sedimentary fill.

These three factors likewise control the form of different geosynclines in the cordillera phase: this includes the degree of dismemberment of the geosyncline, the disposition of the cordilleras and intervening furrows, the degree of permanence of these elements, and so on.

THE EVOLUTION OF CORDILLERAS AND FURROWS

The cordilleran phase is characterised by considerable instability. It is in geosynclines which have reached this stage of development that the epicentres of earthquakes of greatest magnitude are located, notably in those of the circum-Pacific belt. Tangential tensions arising from regional tectonism vary considerably with time, whatever their cause. They do not act continuously in precisely the same direction. Also, they probably effect some modification of the isostatic equilibrium itself. The shifting equilibrium resulting from the opposition of these forces is even more changeable than the forces themselves taken individually, and is, in fact, essentially unstable.

Moreover, the framework of the geosyncline is changing. New fractures are produced which break up the major components, the cordilleras and the furrows. These components are themselves subject to irregular oscillatory movements, sometimes uplift, sometimes downwarping. The study of presentday earth movements in Japan is significant in this respect. Some portion of a geosyncline, delimited by faults, sinks gradually for a period without great earthquakes and then is uplifted sharply during a violent earthquake and is raised to an altitude higher than before the beginning of subsidence. Chile and California display equally convincing evidence. Similarly, horizontal displacements can occur, such as along the great San Andreas fault in California. The combination of both normal and tangential tensional stresses accounts for all types of displacement: horizontal, vertical and oblique. On the whole, however, this instability does not produce the same effects in the mesogeic zone as in the circum-Pacific suite. In the mesogeic belt, volcanic eruptions are relatively rare. It has always been

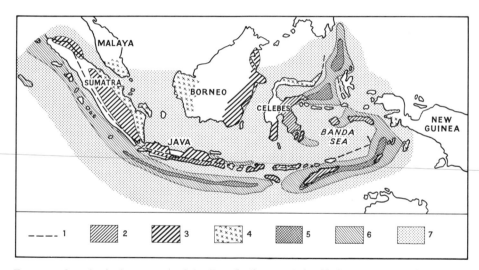

Fig. 2.3. Geophysical structure of the East Indian arc (*after Umbgrove, 1949, and Aubouin, 1961*)

Moulded on the margin of the Malayan platform, the arc comprises successively, from the inner to the outer edges, aligned undulating basins of subsidence, followed by a Miocene folded arc with a line of volcanoes at the contact, followed, finally, by a deep-sea trench with a strongly negative isostatic anomaly indicating a marked tendency to downwarping

1. Volcanic arc
2. Regions folded in Miocene
3. Subsiding areas lightly folded at end of Pliocene
4. Stable blocks. un-folded

Regional Isostatic anomalies:
5. From − 250 to − 100
6. From − 100 to 0
7. From 0 to + 150

so: magmatic effusions are not abundant in sediments of the cordilleran phase, being local outpourings limited both in space and time. On the other hand, in the circum-Pacific belt, vulcanism plays a major role, and is of long-standing importance. Eruptive rock formations abound in the Mesozoic of the Andes of Peru and Chile, as in those of Japan. Many atolls on the island arcs of the western Pacific have a volcanic basement of Tertiary or Cretaceous age.

The cordilleran phase is characterised, above all, by an intense morphogenetic activity. Tectonic instability, in causing elevated blocks to appear between furrows which subside and sometimes allow the growth of important volcanic cones, creates marked irregularities on the surface of the earth: this provides great potential for landform development. Erosion is readily effected in these narrow mountainous zones, close to the base level provided by sea level. Earthquakes contribute to the instability of slopes. This accidented relief favours violent orographic rainstorms. Japan, for example, is one region of the earth where morphogenesis is most actively in progress at the present. Dissection progresses rapidly under the action of short torrents on severe gradients, draining very steep slopes and discharging rapidly into the sea. The debris derived from erosion is readily transported and builds up into deltaic cones which decline by way of a subaquatic scree slope to the bottom of the neighbouring furrow.

Tectonic instability stimulates abnormal rates of landform evolution. The supply of debris is most rapid and severe during the preceding phase. Each furrow receives the material eroded from the neighbouring cordilleras. Distances being short, little is lost *en route*. Sedimentation is not widespread. The materials, violently transported and rapidly deposited, have little time in which to become modified: often they are poorly graded. Subsidence, submarine slumps, and turbidity currents perpetuate the torrential transportation of debris, and yield detrital sediments which are a mixture of pebbles, sands, clays and indeed even blocks of several cubic metres. This is the *Wildflysch* facies of the Eocene beds of the Alps. Further away from the cordilleras or in less disturbed zones, is found the graded bedding of turbidity currents but generally in much thinner beds, of only a few centimetres in thickness, emplaced by less abundant but frequently repeated flows. This is the *flysch* facies.

At great depth, disruption of the geosynclinal trench facilitates magmatic outbursts. As we have seen, some of these take the form of subaerial vulcanism. But this is not the only form. Outpourings occur also beneath the sea, within the furrows. They give rise particularly to some special volcanic rocks, greenstones and ophiolites, like those of Eocene age which run through the *schistes lustrés* of the Queyras. The invasion of sedimentary rocks by magmatic minerals, moreover, gives rise to certain types of metamorphism. For example, the Alpine *schistes lustrés* are a flysch facies, deposited in a geosynclinal furrow and very slightly metamorphosed. Crystals of sericite, parent mineral of the micas, are formed particularly between the flakes of sedimentary shales, quartz veins having filled the fissures and calcareous beds having recrystallised to calcareous slate or schist.

These igneous occurrences, together with the mechanical effects of tectonism, are of major importance. The material is modified, becoming denser and more rigid, and so contributing to a certain relative stability. The tendency for the geosyncline to break up is diminished. It is tempting to say that sections of the geosyncline are partly welded together again. A third phase of development is thus initiated, that of progressive stabilisation.

Tendency towards gradual stabilisation

The intense morphogenetic activity which characterises the cordilleran phase increases the regional isostatic imbalance of the geosyncline. In effect, the denudation of the emergent land masses rapidly augments the volume of fresh sediments of low density which accumulate in the furrows. In aggregate, the volume of these fresh sediments is greater than 20 to 30 per cent of the volume of the source rocks from which they are derived. The unburdening of the eroded cordilleras and the sedimentary loading of the furrows create movements of isostatic type, as has been shown by Gourinard's studies of Quaternary deformations in the vicinity of Oran. These movements tend towards a progressively greater stability, while at depth the igneous invasions increase the rigidity of the basement in metamorphosing the sediments and in filling fractures and fissures. Both series of phenomena act in the same direction and increase with time. Finally, they overcome the regional tension and the tendency to disruption engendered by their interference with isostatic readjustments. During the stabilisation of the geosyncline tensional effects are often carried forward towards the borders, i.e. towards less well-developed areas where the processes of stabilisation are less advanced. A new geosynclinal trench then forms along the margins of the older geosyncline. Such coupled geosynclines occur quite frequently. The Alpine geosyncline is bordered on its northwestern side by regions which constituted part of the Hercynian geosyncline in Europe and which mould themselves around the margins of the Fenno-Scandian shield. Similar relationships occur in central Asia and in Australia (Fig. 2.4).

This tendency towards greater stability is slow-moving and lasts for a long time, from 50 to 70 million years: almost as long on average as the two preceding phases taken together. This progression is far from being uniform and regular: often there is some reversal of the process in the form of episodes of growing instability, of tectonic paroxysms lasting up to some millions of years. In these paroxysms, regional tension seems to play a major role. The majority of them appear to have been reactivated by the isostatic imbalance produced by the first glaciations: in most of the folded mountain chains of the globe, in fact, a tendency to vertical movement with

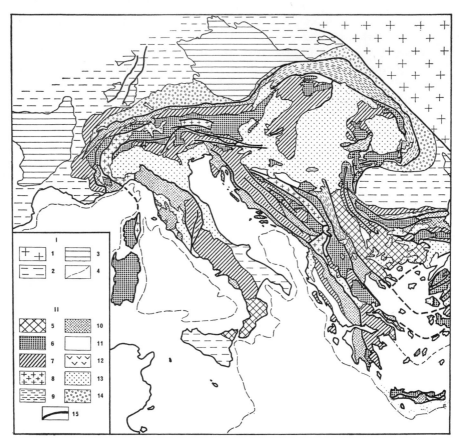

FIG. 2.4. Schematic map of the tectonic evolution of the central Mediterranean (*after Muratov, 1962, p. 188*)

1. Pre-cambrian Russian platform
2. Hercynian platform of western Europe, covered by later sedimentary formations
3. Granitised Hercynian massifs of western Europe
4. Outline of submarine basins
5. Folded Palaeozoic massifs
6. Palaeozoic cores of anticlinoria
7. Thin Mesozoic cover, mostly limestone, on folded Palaeozoic basement
8. Geosynclines with Triassic and Jurassic rocks, including volcanics
9. Geosynclines filled with rocks from late Jurassic to late Palaeogene
10. Geosynclines filled with flysch
11. Geosynclines of late Eocene and Oligocene period
12. Vulcanism in geosynclinal troughs of Jurassic, Cretaceous and Palaeogene
13. Intra-montane basins
14. Piedmont accumulations
15. Major fault zones

The map shows the major tectonic regions, in particular the successive positions of the geosynclinal troughs and the principal fold belts

uplift of some portions and downwarping of others can be discerned toward the end of the Pliocene and in the older Quaternary.

The tendency toward gradual stabilisation is characterised in the geosynclinal zone by the predominance of uplift which constitutes a reversal of the general tendency observed in the preceding phases. Local cordilleran uplifts tend to become general and to extend themselves throughout the whole geosynclinal zone. For this reason, and despite the fact that the cordilleras are already elevated into mountains, it is permissible to use the old term orogenesis for this stage. Subsequently, the earth movements cease progressively and isostatic readjustments become predominant.

THE OROGENIC PHASE

This stage is characterised by important and generally widespread uplift. The whole geosynclinal ensemble emerges. The sea is pushed back toward the margins of the zone and occupies piedmont regions, such as those of the molasse at the foot of the Miocene Alps. The result is a mountain chain, generally massive and of varying altitude, which differs from the alternating furrows and cordilleras of the preceding phase, even though certain of its components are less elevated than their neighbours and form depressions within the chain. The western Alps are at this stage of their evolution, although almost toward the end of it.

The mechanisms which set in motion the orogenic phase are poorly known, and seem to vary quite markedly according to whether one considers the mesogeic zone or the circum-Pacific belt. In both cases, deep-seated metamorphism has been invoked. Granite, of course, has a lower density than the basalts. The transformation at depth within the geosycline of sedimentary rocks into gneisses and granites instigates an isostatic readjustment in the form of uplift of the whole geosynclinal tract. One might even go so far as to compare the granite masses to diapirs. The more intense granitisation of the geosynclines alone differentiates them, therefore, from the ancient massifs and continental shields. While the orogenic mechanism is probably acceptable, it seems inadequate on its own to account for all known facts: it has been thought necessary to invoke sub-crustal currents which must, moreover, play a part in granitisation.

Nevertheless, regional movements also occur and certain limited displacements of the continental blocks must be admitted in order to account for the shortening of strata during folding. However, as we have already seen, this cannot be accepted as a general explanation of the origin of the continents. The tangential components of the regional earth movements have, moreover, different effects in each of the earth's two great mountain types:

1. In the mesogeic zone, it is generally agreed that effects of compression are the result of the convergence, in certain geological epochs of continental forelands situated on either side of geosynclines (Africa and Europe, Russo-

Siberian platform and Arabian shield, central Asian platform and the Deccan shield). Such continental convergence served to narrow the geosynclines: it also contributed, simultaneously, to folding and uplift, the resistance offered by the atmosphere and the hydrosphere being less than at great depth on the margins of the sial. Conversely, during other epochs, the continental forelands have had a tendency to separate one from the other, such that tensional stress and associated rifting and downwarping has prevailed. This has been the case in the western basin of the Mediterranean since the Miocene.

2. In the circum-Pacific belt conditions are different because the chains are unilateral. They border the continental blocks and descend towards the ocean basins. Tangential movements occur: the San Andreas dislocation is one expression of this. Some authors go so far as to suggest a kind of rotation of the whole Pacific basin. But the results of these displacements are not the same as in the mesogeic assemblage: they cannot be compared to the jaws of a vice! They do not give rise to compression effects.

It follows that the two tectonic styles are different. In the mesogeic assemblage, notably all around the Mediterranean where the effects of compression have been most intense, tangential displacements are particularly important. Beds are severely contorted and overfolded. *Décollements* (zones of detachment of cover rocks and basement) occur either along the more plastic horizons, or along shear planes. They frequently disturb the continuity of the strata, yielding splinters when the displacement is weak (some kilometres at the maximum), or overthrust nappes when it exceeds ten kilometres and when the site of the dislocation (the root of the nappe) is difficult to determine. Until recent years, everyone agreed that nappes resulted from disturbances of rock masses to some depth. Recently, the theory of gravity tectonics has been advanced, according to which nappes are viewed as mountain sides which become detached from their substratum and glide down the flank of the orogen from an elevated portion towards a less elevated area. Such *décollements* certainly exist, for example on the limbs of tight recumbent folds which have generated during their formation substantial voids. They give rise to stretched and dislocated strata, known as *collapse structures*. At the foot of the anticlinal limb where they originate, collapse structures form accumulations of shattered and chaotic rock masses piled one upon another. All this is very different from the overthrust nappes, from which they are distinguished by their small dimensions, by the steep slopes necessary for their formation, and in most cases by the very nature of the strata themselves. It is doubtful whether nappes can be put into motion under the influence of gravity alone on slopes averaging only a few degrees. Otherwise, no deep gorges would ever be produced: as in certain clays, they would become filled up progressively by the flow of debris. It is also unlikely, as for example in the case of the nappes of the Pennide Alps, that masses of granite and gneiss were overwhelmed in such glissades or, as in the Helvetic nappes, that rock wedges injected the autochthonous formations, as is the case in involuted nappes. The theory of gravity tectonics seems to have resulted from a confusion of scale: it is rather improbable. Apart from collapse structures, which are another consideration, the idea of gravity tectonics rests on no conclusive observation, on no precise and weighty reconstruction; moreover it is not in the course of the orogenic phase that the most favourable conditions exist for it, but rather in the cordilleran phase. It is then, in fact, that tectonic disruption generates great variations in level over short distances, and repeated earthquake shocks facilitate gravity gliding. The detachment of whole sequences of beds from the flanks of the cordilleras and their sliding towards the neighbouring furrow where they give rise to collapse structures is probable at this time. But it cannot act over distances of some hundreds of kilometres, as has been admitted by the 'nappistes' in the Alps, the south of Spain, and the Rif or the Tell of Algeria. Nevertheless, even if (as in my opinion) we must minimise their amplitude, these tangential displacements do exist and are a characteristic of the mesogeic assemblage which surrounds the Mediterranean.

Tangential displacements are much less important and less widespread in the tectonic regime of the unilateral chains of the circum-Pacific zone.

FIG. 2.5. Collapse structure: the 'Barre de Sisteron'
Upper Jurassic limestones inclined vertically on the southern flank of the Sisteron dome, and forming an abrupt and narrow ridge in between marl outcrops that underlie and overlie the limestones. The limestone beds, when they were still pliable, slid down the flank of the up-warping dome and are crumpled on themselves. This is well seen in the core of the ridge, whilst the beds on the two sides have a more normal disposition

In Japan as in Peru and Chile, it is the vertical movement which dominates, resulting in differentially uplifted blocks, either along the length of the major faults as in the central Chilean depression, or in the form of major upwarps. These are possible even in the rigid rock masses such as the Peruvian coastal batholith, made up of granite, thanks to innumerable small fissures which have allowed movement of the whole assemblage. The main tectonic stresses are vertical. Folding is no less in evidence although it is subordinate. Many of these faults are obliquely aligned and often contain an important horizontal component. A trench bounded by faults converging at depth effectively foreshortens the beds which are subsiding within it. Such conditions are common. We have, then, usually moderately folded beds within the trench, itself lying between two tabular horsts. On the planes of transverse faults having major lateral displacements such as the San Andreas "fault", pressures develop which give rise to folded structures. Near-surface tectonics sometimes give rise to fault splinters and overthrusts, as in the regions of folded cover rocks. In California, Peru and Chile folding is always present locally, on certain fault-bounded tectonic units, the faults owing their origin to tangential movements. Nowhere do the *décollement* folds of the cover rocks, and even less the overthrust blocks, approach the importance they possess in the Mediterranean regions, despite the intensity of the tectonic movements and a more imposing scale of orogenesis. The absence of the effects of compressive folding on a regional scale, resulting from the unilateral disposition of the chains, seems to be the cause.

Step by step throughout this phase, the differential movements decrease and movements of the whole assemblage over an increasing radius become more and more predominant.

DEFORMATION OF THE GEOSYNCLINAL SUITE

The passage from the cordilleran to the orogenic phase is one aspect of the growing importance of the role of deformation of the whole geosynclinal assemblage. Instead of unstable strips, some foundering in geosynclinal furrows while others are uplifted into cordilleras, the whole geosynclinal suite is upwarped and makes up a gigantic *geotumour* or *undation*, to use the terminology of some Germanic structural geologists. This updoming is accentuated throughout the stabilisation phase. It continues to act even after the completion of the folding and scarcely modifies it. Orogenesis continues, therefore, at a time when tectogenesis (particularly intense during the cordilleran phase and at the beginning of orogenesis) is practically finished. For example, tectogenesis in the northern Alps was substantially completed during the Miocene, yet an important uplift of the massif occurred at the end of the Miocene, in the Pliocene and in the Quaternary. This has given the Alpine chain its symmetrically updomed appearance.

Throughout this period of general upwarping, uplift is uniform neither in space nor in time. Migration of axes of uplift and differential movements are common. Migration of axes of uplift is the most important phenomenon in a chain such as the Alps. In the most arcuate part of the chain in Switzerland, Italy and France, there is a notable displacement of these axes of uplift towards the margins of the arc. In the pre-Alps, for example, a faulted platform structure, having given rise to *plis de couverture* in the Eocene, developed a tendency to downwarping toward the end of Oligocene which became even more marked in the Miocene. In the early part of the orogenic phase, it was transformed into a piedmont and enormous masses of debris, eroded from the adjacent developing upwarp, gave rise to the *molasse* made up of sands variously argillaceous, calcareous and gritty, passing into gravels and conglomerates, laid down at various times and in various localities within marshes, debris cones (*cônes de déjection*), deltas, lakes or even arms of the sea. In geological parlance, the word 'molasse' designates those types of deposit accumulated on a piedmont during the orogenic phase, while the flysch deposits, as we have seen, correspond to accumulations of geosynclinal furrows derived from the cordilleran phase. Their respective places in the evolution of the geosyncline differ, although in certain cases their facies are similar. The pre-Alpine molasse was subsequently overwhelmed in the uplift at the end of the Miocene and affected by the folding of the cover rocks as may be clearly seen in the Grande Chartreuse. The zone of downwarping was also carried forward beyond the limits of the chain, in the region of Lake Geneva and on the Rhodanian Furrow. In the view of the proponents of gravity tectonics, *décollement* in the cover rocks producing nappes may occur at the time of uplift. Thus, the massifs of the pre-Alps have been interpreted as the remains of the cover rocks of the central massifs (e.g. Belledonne for the Grande Chartreuse), which slid towards the outer portion of the mountain arc during the post-molasse uplift. The Alpine Furrow occupied the shatter zone opened up by this downsliding. The tectogenesis, according to this viewpoint, would continue throughout the whole of the orogenic period, seeing that this provides its motivation.

While in France folding extends to the outer margins of the mountain arc, in Italy it has been reduced by the foundering around the edges of the Po lowland. In Lombardy and in the Piedmont, the mountains rise sharply from the plain. Great fault scarps cut across earlier folded structures. The sedimentary cover, which is largely preserved in France and in Switzerland, has practically disappeared in this region. The central crystalline massifs of Hercynian age are close to the margin. The Miocene and Pliocene molasse sediments were covered rapidly by about a thousand metres of Quaternary material. The drainage network has not yet had sufficient time to adjust to the broad area of subsidence: the watershed remains very close to the Po lowland.

These characteristics, which together have imposed a marked asymmetry on the mountain chain, are restricted to the most arcuate segment: they disappear as one moves eastward, in the Dolomites and in Austria. There the essential characteristics are associated with differential movement of blocks. This is a very general condition. It is the arcuate mountain chains which always provide the clearest examples of asymmetry and of migration of axes of uplift toward the convex margins of the arc. It is as though, during the stabilisation phase which began with orogenesis, the chain was forced to increase its radius of curvature.

Block movements are particularly apparent in the straight mountain chains. This is true in the case of the Pyrenees, the eastern Alps, and the Chilean and Peruvian Andes. They combine with the general upwarping to which they are subordinate. In the Pyrenees and the Alps, they are particularly characteristic of the eastern extremity of the mountain chains, which literally taper toward the edge of two vast sunklands, the Gulf of Lyon and the Pannonian Plain respectively. Horsts stand out between grabens, such as the plain of Roussillon at the foot of the Albères and the Canigou blocks. Block movements have occurred throughout the Quaternary as much in the eastern Pyrenees as in the eastern Alps. Vigorous erosion of the uplifted blocks has furnished the material of clastic sedimentation, either marine (Pliocene of the Rousillon), lacustrine (Sarmatian (Miocene) beds of the Pannonian Basin), fluvial or fluvioglacial, large cones developing in the depressions where subsidence favours the spread of floodwaters.

Block movements of gigantic proportions have occurred in the Andes. In Peru, the general appearance of the upwarping is well seen in the coastal granitic batholith, which rises slowly and almost uniformly from the Pacific to almost 3 000 m above sea level near Lima or Nazca. Further south, towards Arequipa and Tacna, the same characteristics may be noted in the Mio-Pliocene volcanic series. Locally, for example at Tacna or Chiclayo, some fault blocks have collapsed during uplift and, in contrast to the broad downwarps, have formed small tectonic trenches where clastic formations of Quaternary age have accumulated to a thickness of some tens of metres. In southern Peru and in Bolivia, the region about Lake Titicaca is a different case: it has taken part in the general uplift but, nevertheless, it remains less elevated than its neighbouring areas allowing it to develop an endoreic basin. It is not unknown for some of the more weakly uplifted sections in the interior of a folded assemblage to be made up of rigid fragments of continental platform material. This is the case on the Spanish Meseta, large sections of Anatolia, Tibet, and the Colorado Plateau. Having some of the characteristics of platform sedimentation, the cover rocks of the shield remain tabular, sometimes being cut by faults and buried under lava (Columbia Plateau, on the United States–Canada border).

In central Chile, between the coastal mountains and the main Cordillera, the Central Depression is made up of parallel trenches which are depressed while the intervening blocks are uplifted. Occupied at one time by lakes and at another time by plains liable to floods, they have received the clastic material eroded mainly from the most elevated mountains, namely the High Cordillera. Such *intermontane basins* are common in the orogenic phase. They develop the more readily as general upwarping produces tensional stress which, in turn, causes subsidence along lines of weakness. In the case of central Chile, the fractures are so developed that abundant volcanic outbursts have occurred at all stages. It may be noted, in cases of this type, that there are certain similarities to some platform areas affected by violent tectonic stress. For example, during the middle and upper Tertiary, the Massif Central was associated with the development of the downwarps of the Limagne and the Forez and the volcanism of the Velay and the Auvergne. These similarities are not fortuitous: they mark the end of the evolution of the geosyncline.

THE END PHASE OF GEOSYNCLINAL DEVELOPMENT AND REGENERATED GEOSYNCLINES

The end of the geosynclinal sequence is characterised by progressive slowing down of orogenesis and related earth movements. There are two sets of reasons for this:

1. Owing to the uplift of the orogenic phase, which lasts many tens of millions of years, the erosion suffered by the geosynclinal mountain chain is considerable. According to Gabert, in the drainage trap formed partly by the Alps and partly by the Apennines, the Quaternary accumulations of the Piedmont represent an average downwasting of 232 m. Such an amount does not seem particularly exceptional: it has almost certainly been surpassed in other mountain chains, such as the Himalayas, and those of Burma, Japan, etc. In the course of an evolution lasting 10 to 20 million years, beds totalling thousands of metres in thickness are uplifted. The deepseated formations are laid bare: they are metamorphic or crystalline. Thus Cretaceous granites and perhaps even granites of early Eocene age have been widely exposed in Chile and in China. Their mechanical properties are not those of sedimentary strata: they are more massive, less plastic, more resistant to folding and, above all, liable to large-scale deformation. If not, they fracture and break up into a series of fault blocks. As erosion, triggered by uplift, removes the more supple superficial strata, the earth movement changes in type and takes on the character of deepseated fold tectonics (*plis de fond*), involving larger and essentially rigid units. Block movement replaces folding in the normal sense. Thus, the situation approaches that of shield areas which have been fractured under severe stress and also display block movement.

2. This major erosion affecting the orogen necessarily stimulates certain isostatic readjustments. The resulting unloading provokes a tendency to uplift, which tends further to maintain the orogenesis. This mechanism may help to explain the relatively long duration of this phase. But the isostatic reaction, which is essentially vertical, acts on its own, particularly on units which are sufficiently large. It tends to give rise to deformation of the whole ensemble, to up-arch the orogen so that downcutting, given sufficient

retrogressive erosion, is at a maximum in the more elevated parts. In other words, isostasy tends to maintain the general character of the undation. Correlatively, the sedimentary overloading imposed on the piedmonts by the accumulation of eroded debris tends to depress them. The 2 000–3 000 m of molasse which is encountered in Switzerland and in Bavaria, where the facies are throughout of very shallow type, implies such a slow downwarping. It is possible that currents at depth displace some of the piedmont material towards the undation, which would maintain the movement. But such mechanisms are operative at the scale of a whole mountain chain and express themselves as movements of the whole assemblage. They do not exclude, as we have seen, movement of adjacent elements along local lines of weakness. But as the deep structures, consolidated by metamorphism, are progressively brought closer to the surface, these weaknesses tend to disappear. Earth movements of large radius (*mouvements à grand rayon*), analogous to those which occur on the platforms, become more and more dominant.

The lithological modifications resulting from the increasing importance of denudation, and the correlative change in the style of earth movement, transform the geosynclinal regions, toward the end of their development, into platform regions. Increasingly, they are characterised by the exposure of metamorphic and crystalline rocks and affected by large scale earth movements in which isostasy plays an important role. The residual relief

FIG. 2.6. Accordant summit levels (*gipfelfluhr*) of the Sierra de la Culata, in the Venezuelan Andes. A block of gneiss uplifted in the Plio-Pleistocene above the valley in the middle distance. The earlier Neogene planation has been sharply dissected as a result of the uplift

which is inherited is progressively reduced. The tectonic depressions, which take the form of basins, are progressively infilled by the clastic materials furnished by the residual relief. Such a picture is provided in Europe by the remains of the Hercynian mountain chains in the Permian, a period during which clastic sediments, from sandy clays to gravels, gradually infilled (sometimes up to several tens of metres in thickness) basins lying between ridges and the remnants of mountains mostly composed of crystalline rocks. This was particularly the case in the region of Saint Dié. Even later (in the lower Triassic) on an even more reduced relief, the first beds of the platform phase of sedimentation were laid down, in the form of the Vosgian Grit, made up of the widely distributed deposits of the littoral plain. The stabilisation of the ancient geosyncline was then achieved. But certain lines of weakness, not completely sealed, can become activated anew during platform development and so guide tectonic events (fractures, undulations, flexures). This is *posthumous tectonism*, important in the sedimentary basins and in the ancient massifs of Hercynian Europe.

This type of development is normal in geosynclines. It explains why the platforms are progressively built up at the expense of ancient fold chains bordering their oldest elements, the shields. At the same time, this evolution is interrupted from time to time. Following the Russian terminology, one may speak of *regenerated geosynclines*.

Regenerated geosynclines result from violent regional tensions affecting a geosyncline in course of stabilisation. The Rockies, in the United States, provide a good example, as do certain chains in central Asia on the frontier of Tibet and Turkestan. Because of the rigidity of deeply eroded synclinal structures, the tectonic style is for the most part of the fracture type. It stimulates movement of blocks, certain elements becoming uplifted and giving rise to massive mountain chains, while others are depressed forming basins, such as the Wyoming basin, and receive detrital sediments. The uplifted blocks are often localised in those areas where the sedimentary cover has already been almost entirely eroded and where the crystalline basement is exposed, while the basins, on the contrary, are generally located in regions previously less elevated where this cover has persisted more widely. The flanks of the blocks, in this case, are characterised by faults or flexures. Therefore the architecture of these regenerated geosynclines owes nothing to the original. It is analogous to the case of orogens fragmented by faults, as in the Alps or the eastern Pyrenees. Essentially, the difference rests in the order of their geological development. Regenerated geosynclines may often become covered by transgressive platform sediments before being made into mountainous regions again: this is not the case with orogens which are beginning to stabilise themselves.

As a general rule, therefore, there is some interference between tectonic phenomena and external activity throughout the evolution of folded chains. But important though they be, the external or epigene processes are always subordinate to those of internal or hypogene type. This is why it is important to stress the activity of the internal forces: failing this, one is bound to understand little about relief development and to commit gross errors on this subject perpetuating sterile ideas such as the conception of the cycle of erosion. Only now that we have outlined the main characteristics of the structural evolution of geosynclines can we approach a study of the relief forms which develop on them.

Relief forms of geosynclines

One important conclusion emerges from this outline of the evolution of geosynclines: *geosynclinal areas are characterised by the predominance of forms due to tectonism.* There are two reasons for this:

1. Tectonic activity is very vigorous and creates a framework within which the subaerial forces have little scope to operate independently, because the modification of the geosyncline is continual and too rapid. We have seen, for example, that despite the advantages of a greatly depressed base level, the torrential Po tributaries have not succeeded in altering the line of the major watersheds to any great extent, their location remaining a function of tectonic asymmetry.

2. The very conditions of sedimentation minimise the likelihood of strong lithological control during dissection. Geosynclinal sedimentation is characterised by a relative uniformity, and by a differentiation of facies less marked than in the case of sedimentation on platform areas. Clastic sediments dominate, only their calibre varying; they range from clays to argillites, to sands, and even to pebbles and boulders. While graded beds very often give rise to an alternation of a wide range of grain sizes within a succession of beds scarcely more than a few metres thick, the scale is too small for the differences to be exploited by erosion, apart from the sculpture by pluvial action along rock surfaces.

Limited lithological contrasts on the one side and vigour and permanence of active tectonic influences on the other, combine to produce a morphogenesis which is closely controlled by tectonics. The areas of subsidence become almost automatically places of marine, lacustrine or subaerial sedimentation. The uplifted blocks are subject to vigorous dissection, little influenced by a lithology which is too uniform at medium scales. The very vigour of this dissection serves to reduce the lithological influence. At first sight, for example, it is sometimes difficult to know whether a particular landscape made up of sharp crests, narrow valleys and long steep slopes is developed in severely crushed granites or in flysch.

We are thinking here on the scale of a massif, a scale which corresponds to the structural components in the interior of geosynclines and therefore, in broad terms, dimensions of tens perhaps hundreds of kilometres. The subordinate features, including folds, thrusts and monoclines will be studied further, because they influence the relief especially when they occur in the cover rocks. As tectonics play such a determining role, the relief of massifs and intermontane basins will be examined according to the structural criteria which we have defined.

At the same time, from the geomorphological point of view, some re-grouping is possible and this will help to simplify our explanation. Two major groups of massifs may be distinguished: those which are 'rooted', in which vertical movements predominate (cordilleras, deepseated folds); and those in which the strata have been subjected to some major faulting, resulting in a tectonic style which is ejective (nappes, klippen, and fault blocks). We deal first with the intermontane basins and then pass on to the general disposition of the relief in geosynclinal chains.

Massifs dominated by vertical movements: cordilleras, deep-seated folds (*plis de fond*)

These vertical movements play a major role throughout geosynclinal development, from the beginning of the downwarping initiating the geosyncline until its final stabilisation. Also, the massifs at some stages of this evolution appear markedly different one from another in their essential characteristics. Nevertheless, they possess enough in common for the development of the relief to offer important similarities.

GENERAL CHARACTERISTICS

These result from structural similarities and are composed of the following:

Vigour of dissection

This results from a structure consisting of large wedges, structural compartments affected essentially by vertical movements, in the present case one of uplift. This disposition creates a violent contrast between the uplifted massif and the neighbouring downwarped components, geosynclinal furrows or intermontane basins. The contact between the two is emphasised by the presence of severe breaks, either faults or folds, which are faithfully reflected in the relief because they are so active. For example, the devastating earthquake of Agadir resulted from renewed activity of the sub-Atlas belt of weakness, the great fault separating the basin of the Sous from the massif of the High Atlas. The steep slopes by which the Ligurian and the Niçoise Alps descend to the Mediterranean correspond to a flank of a cordillera. In mountain chains of *plis de fond* or cordilleran type, the margins of these wedges always form a sharp tectonic escarpment. This escarpment, in fact, is due directly to the differential movement of two adjacent structural compartments.

Such severe differential uplift engenders considerable morphogenetic potential. It increases the intensity of the gravitational processes. Torrents rush down the slopes and readily transport the debris, even that of the coarsest calibre. The steep slopes favour screes, landslides and slips of all kinds. The rocks themselves sometimes yield and become detached (*décolle-ment*), producing collapse structures. In fact, such marginal slopes of the wedges of massifs are particularly suitable to the development of collapse structures. This is true of the alternating sandstone and marl strata on the northwestern margins of the trench of Lagunillas in the Andes of Venezuela, where the escarpment has grown by at least a thousand metres since the early Quaternary.

The edges of the marginal wedges of the massif also form a fringe of morphogenetic activity constantly renewed, due to continual tectonic movements. The tremors released by them, moreover, favour the development of screes, landslides and even collapse structures. Where tectonic activity is intense, as on the margins of the Lagunillas trench, the rocks appear crushed and fragmented in outcrop and it is often difficult to decide if one is in the presence of bedrock *in situ* or of slipped material or scree. Such fragmentation of bedrock is, from all the evidence, conducive to erosion and allows the retreat of escarpments between the incised scarp front torrents. Even when the faults are vertically disposed, such escarpments never present a steep, uniform face, because the rocks of which they are composed are too readily comminuted. Their slopes are often steeper than the maximum gravity slope for rubble (about 45 degrees), which renders them unstable and liable to become mobile during earthquake shocks. Thus, at Lagunillas, the average slope approximates 60 degrees on many ridge ends.

FIG. 2.7. The rift valley of Lagunillas, in the Venezuelan Andes
The southern edge of the rift valley, formed by an uplifted block of schist, bearing traces of Pliocene planation. Within the rift, the masses of detritus have been compartmentalised by minor fractures.

Due to tectonic movements, torrential streams are generally little incised into the margins of the fault-bound wedges when they are fully active. In effect, the slope grows continuously as a result of the active deformation, and stream incision succeeds in reducing it only with difficulty, more especially as the stream is choked with the abundant debris supplied to it. When the escarpment is tectonically very active and the rocks so resistant to incision, as at Lagunillas, there develops a series of parallel gulleys, weakly incised, showing little ramification, and with a very steep longitudinal profile.

When incision is easier, as for example in shales or in shaly flysch, the gradient of the streams is lower and dissection is more readily accomplished within the fault block. In this case the valleys are more incised and the escarpment is dissected into narrow ridges, of razor back form, their flanks swept periodically by slides. All parts of this landscape of dissection depend on the relative speed of uplift on the one hand and the rate of incision of talwegs on the other. Further, the latter is controlled by the lithology. Therefore, it varies from one type of structural situation to another.

The weakness of differential erosion

Lithology is a major factor especially as it is expressed in the rate of stream incision. But conditions are generally rather unfavourable for the development of a relief due to differential erosion. There are two main reasons for this:

1. The frequent tectonic fragmentation of the rocks, which tends to reduce the range of morphogenetic response. In fact, the more compact of the coherent rocks turn out to be the most shattered, the most broken. Reduced to fragments readily moved on steep slopes, they cease then to behave as 'hard' rock for they no longer require a preparatory phase of fragmentation. The alternations of grits, sometimes even quartzites, with clays in the escarpment of Lagunillas is not expressed in the landforms, the strata being dislocated in great blocks and rubble which becomes mixed with the clay and slides over it.

2. The monotony of geosynclinal facies, excluding lithological contrasts in all typical series of turbidites and flysch, whether metamorphosed or not. Heterogeneity of detail, at the scale of a single stratum or thin bed tends to merge, at the scale of subaerial dissection, into a gross homogeneity which excludes differential erosion.

Accordingly, lithological differences, often reduced by tectonic fragmentation, play a small part. The slopes become uniformly steep as soon as the streams are incised to any extent and lithology is thus able to play a part, notably through the medium of the geomorphic processes. Mudflows, landslips, and torrential water movement are more frequent where shales release more clay; in contrast, scree and rock slides predominate when shales are absent. Likewise, tectonic fragmentation of coherent rocks facilitates stream incision, and, even in tropical regions, differential erosion is little developed. Relatively few waterfalls or rapids occur as a result of streams crossing resistant strata.

Nevertheless, these general characteristics are more or less well expressed as a function of the structural development. The spreading of lavas during a volcanic episode following tectonic fragmentation of the substratum, can give rise to severe lithological contrasts. Such is the case near Lima in Peru: the granites of the coastal batholith, completely fragmented, are folded into a broad anticline and give rise to slopes of 50 to 60 degrees, slopes on which debris fragments move under gravity. At the same time, mud flows of andesite debris give rise to well stratified layers, very little fragmented, and having the appearance of sedimentary strata: these form nearly vertical escarpments. The valley of the Rimac, which is broad and deep on the granites, becomes narrowly confined at the Gorge of the Infernillo. In regenerated geosynclines, on the other hand, the fault blocks are made up of more rigid rocks which have had all the time taken up by the phase of stabilisation of the first geosynclinal cycle in which to become indurated. They are often covered with the remains of sedimentary strata deposited between times and which offer some possibility of differential erosion.

It is now possible for us to examine the characteristic aspects of the various types of structural units.

DISTINCTIVE QUALITIES OF STRUCTURAL UNITS

The tectodynamics and architecture[1] of massifs are functions of the place they occupy in the geosynclinal evolution. Accordingly, let us look at those types defined in our account of the latter.

The relief of the cordilleras

Cordilleras are particularly noted for characteristically rapid uplift of sectors alongside downwarping trenches, which are often invaded contemporaneously by the sea. The sharp structural features, faults and flexures, which delimit them, frequently cut across structural units of earlier date. Such is the case in Japan, in the Niçoise Alps and in Liguria. Also, they display a degree of architectural variety.

Certain cordilleras, such as those in New Zealand around Wellington, show very little lithological variety, being made up of geosynclinal sediments. The part played by differential erosion is here reduced to a minimum. Almost all the relief is tectonic. Differentially warped blocks give rise to massifs with severely dissected crests, whose general appearance faithfully expresses movement of crustal blocks. The summits of the hills, whose altitudes differ little from one to another (accordant summits), rise or fall in steps. The slopes, severely dissected thanks to poor rock resistance and

---

[1] By the word *architecture* is meant the disposition (tectostatics) and the nature (lithology) of the beds or, if you like, the static aspects of structure.

abundant runoff, are all derived from faults and make up escarpments which are more or less degraded along the original line of faulting. The most recent faults give rise to small sharp slopes locally cutting across the general gradient, for the processes of slope development have not yet had sufficient time to obliterate them. This picture presents an extreme case to be found mainly where tectonic influences alone control the relief.

However, tectonic influences are by no means always predominant, as may be judged from a study of the Ligurian mountains made by Fel (1962). These mountains form a cordillera between the Plain of the Po and the Gulf of Genoa, both areas of downwarping. They are still tectonically active at the present, as indicated by earth tremors. This asymmetrical wedge, inclined towards the plain of Lombardy, intersects several tectonic units. Its evolution can be traced back to the Oligocene (30 to 40 million years ago). During this epoch, in fact, an Eocene sedimentation of marine flysch was followed by clastic sedimentation which fossilised a folded relief. Local planations followed, sometimes cutting across the Oligocene gravels. The Oligocene beds are found deposited sometimes in the interior basins and sometimes in the marine trenches. They attain considerable thicknesses and they have fossilised a sharply accidented relief. Their emplacement was accompanied by deformation, generally orientated differently from those of the Eocene folds. Their clastic facies include gravels, marls and sands. This allows the local appearance of exhumed relief, particularly slopes and ridge crests buried in these accumulations. But the major elements of the relief stood above the geosynclinal trenches where these were developing: they were subjected to violent erosion throughout the period of crustal deformation. Here and there local planations, which are always present, have been able to fashion the relief. The planated surfaces have generally been preserved by a marine transgression, which has allowed the deposition of the molasse. Erosion surfaces have continued to develop locally, on the margins of uplands overlooking the sea, particularly in the less resistant formations. Because of insufficient time for formation, they are generally imperfect and are characterised by an irregular landscape. In the Pliocene–Quaternary the whole assemblage was affected by a tectonic paroxysm which gave rise to the present mountain sector with its asymmetry. In the interior of the sector broad folds and faults occur together, with faults predominating. The tectonic style is particularly complex because the cordilleran arc intersects the frontal structural units almost perpendicularly. These have interfered with the recent upheaval. The mountain block is thus composed of a series of elements differing in their architecture: the granite massif of Savone passing, to the SSW, into crystalline schists, the greenstone intrusion of Arenzano, zones of flysch, of *schistes-lustrés* and shales, the Liassic and Triassic calcareous-sandstone cover of the south of the massif of Savone, and finally clastic sediments of Oligocene age which continue to the margins of the Lombardy Plain beneath which they plunge.

The relief is as complex as the structure. Certain summits have been rounded by a prolonged period of erosion, having been under attack since the Oligocene and uplifted in the course of successive earth movements. Others have been exhumed from localised Oligocene infills. They are sometimes very fresh, having been buried before being seriously denuded. Landscapes possessing accordant summits mark local planations variously warped and dissected: however, the conditions favouring such planations have always been rather fleeting. Due to the wedgelike structure, many major slopes border deeply incised valleys. Some of these slopes have been more or less completely exhumed since the Oligocene. Others are recent fault escarpments, while yet others are the remains of erosional slopes which stand above the emergent local planation surfaces in a variety of relationships. The ensemble constitutes a veritable maze when viewed in detail. On the other hand, if one is content merely to observe its general appearance, the predominant influence of recent tectonism appears clearly with the asymmetry clearly impressed on the mountain sector. The movements continue at present, in fact, as is demonstrated by seismic shocks and geophysical anomalies. Because of the recent character of this still incomplete uplift, differential erosion is slight despite the lithological variety. Also, the drainage network retains an asymmetry resulting directly from tectonics: short streams descending towards the Gulf of Genoa have extended their basins at the expense of the longer streams descending towards the Plain of Lombardy in exceptional cases only.

The two examples presented above correspond to the two most widespread types of cordilleras: *New Zealand* type, in which relatively uniform geosynclinal sedimentation, augmented by volcanic episodes, dominates, and where very vigorous dissection is guided solely by tectonism; and *Ligurian* type, where recent tectonism is also all-important and gives to the relief its overall appearance, but where a longer period of evolution results in a variety of architectural units whose influence is discernible in the pattern of dissection and in certain local forms (planation surfaces more or less dissected, exhumed forms and old, residual relief features).

The relief of the deepseated fold zones (plis de fond) *in regenerated geosynclines*

Deepseated fold zones, like cordilleras, are essentially wedgelike structures. This explains their similarities. But, being produced towards the end of the geosynclinal sequence and, principally, in regenerated geosynclines, they affect structures which have already become rigid, most often a crystalline shield and cover rocks made up of non-geosynclinal sediments, marine beds of platform areas or detrital accumulations. Tectonic fragmentation is less. At the scale of a mountain massif, folds developed at depth in regenerated geosynclines constitute a transitional type between the Ligurian type of cordilleras and the horsts of the platform areas subjected to violent tectonic effects, as with the Vosges or the margins of the East African rift valleys.

Their tectonic evolution is maintained, moreover, from one to another as we have shown above.

In the mountains of Central Asia as in Tian Shan, the Sayans or the Pamir and also in the Moroccan High Atlas, the age of the active geosynclinal phase is essentially Palaeozoic. By the middle of the Mesozoic, stabilisation was already well advanced and sedimentary cover rocks of platform type had been emplaced. They have been affected by deepseated folding. But these cover rocks retain, nevertheless, a certain tectonic instability. They do not have the regularity seen in the strata of sedimentary basins affected by slow subsidence. The detrital facies of sandstones and conglomerates, which are supplied by a residual relief affected by earthquake shocks, are frequent. Discordancies multiply, engendered by frequent and important variations in thickness: here the strata thin out into a feather edge, there they pile up in a downwarped trench.

From the architectural point of view, therefore, the deepseated fold zones are characterised by:

1. Less tectonic fragmentation, resulting from their rigidity which, in turn, is due to their old complex structures: this yields wedge-shaped blocks clearly delimited and less fractured than is the case with cordilleras, having essentially moved as a block.
2. The existence of sedimentary cover rocks, originally more or less tabular and not too dislocated, presenting both the lithological contrasts particularly associated with *plis de couverture* and the structural forms which are cut into them. But these folded cover rocks are not typical of platform regions situated on the margins of geosynclines and affected, at their contact, by tangential deformations. Nevertheless, forms of the same general type with dimensions in kilometres or even tens of kilometres can be found.

Certain Hercynian massifs caught up in the Alpine geosyncline present some of these characteristics in part, as in the case of the Pelvoux or the Belledonne. On the eastern flank of the Pelvoux, for example, a cover of Triassic rocks remains attached to the crystalline shield and, being well tilted on an average of thirty degrees, allows the development of a peripheral furrow (valley of Clairée), but the violence of the tectonic movements in the geosyncline has severely shattered these massifs.

The Moroccan High Atlas, systematically studied by Dresch, provides a good example of deepseated folding within a regenerated geosyncline. Its evolution is characterised by the following stages:

1. Folding in the Palaeozoic of a series of limestones, schists and gritstones, with igneous intrusions; in the Carboniferous, Permian and Triassic, a substantial series of sandstones, argillites and evaporites was deposited in basins during a period of stabilisation; then a period of surface erosion

truncated the whole, marking the completion of this period of geosynclinal development.

2. A sedimentary cover of platform type fossilised this Triassic erosion surface. In the Jurassic, conglomerates, marls, limestones, and grits were deposited with intercalations of basalt. In the Cretaceous the marine transgression was accentuated, with marine facies in the west and continental facies in the east.

3. At the end of the Cretaceous, the marine sedimentation was interrupted by new tectonic deformations. An erosion surface developed in the Eocene, was deformed in its turn, here dissected and there buried under detrital accumulations of Oligo-Miocene age.

4. Very important block movements occurred towards the end of the Miocene, giving rise to the present mountains.

The tectonism giving rise to these mountains is characterised by recumbent folds, flexures, faults, and fault-folds (*plis-failles*). Sometimes small overthrusts may be observed along the line of oblique faults, but their amplitude does not exceed a kilometre and there are neither imbricated blocks nor klippen and even less major overthrusting. The tectonic style differs from one region to another as a function of the behaviour of the material: broad domes and basins affect the sedimentary cover where it is thickest, although there is no *décollement*; folds appear in the shales towards the west; the granite massifs and the lava accumulations are particularly cut by faults.

The relief forms which develop within this architectural setting, controlled by the dissection stimulated by uplift of the main mountain arch, are as follows:

1. Parts of more or less deformed erosion surfaces are, for one thing, exhumed from beneath different cover rocks. This can be seen mainly on the margins of the massif where the edges of the mountain wedge do not plunge too steeply. This association of forms resembles that observed on the margins of many of the ancient massifs: elements of surface erosion exhumed where hard rocks are truncated, and homoclines carved in the cover rocks. Other remains of erosion surfaces appear in the interior of the structural wedge in the form of hilltop levels more or less dissected, some low with respect to others due to unequal uplift arising from block movement in the interior of the wedge.

2. Some gorgelike valleys incise the currently rising mountain wedge, as is shown by the Pliocene–Quaternary deformation of the margins and by certain active faults. As the materials in such an assemblage are resistant, like those in the ancient massifs of the platform areas, incision of the talwegs is slow and difficult. Retrogressive erosion penetrates but slowly into the interior of the massif, facilitating the persistence of planation surfaces. Slopes remain steep and very resistant, hence the gorges.

3. Structural forms resulting from differential erosion. Their persistence,

due to resistant lithology, is facilitated by the long-term permanence of the uplift and by the slowing down of the dissection due to the resistance offered by the basement. This gradual dissection brings out the importance of lithological differences, in a different way from the situation in the cordilleras. In the undulating strata of the cover rocks, a collection of Jura-type forms is encountered: anticlinal valleys and perched synclines. In the more rigid beds of the Carboniferous, Permian and the Triassic, homo-clinal escarpments and plateaus of tilted blocks, bounded by faults, dominate. Finally, on the basement, an Appalachian relief is developed with excavation along schistose bands.

4. Faults, the principal structural feature of block tectonics, are important throughout, but their ages vary. Some are active, while others became active in the middle of the Tertiary or even earlier. Also, the escarpments present every stage of development down to the final stage of levelling. In the crystalline parts of the shield where lithological differences are less marked, variously degraded fault escarpments constitute the principal forms, associated with the remains of more or less dissected planation surfaces.

The variety of architectural conditions, the length of the period of dissection and the limitation of this dissection by rock resistance are, therefore, at the source of the varied relief forms, more varied than in the cordilleras; they give to the deepseated fold zones of the regenerated geosynclines a place which is intermediate between the cordilleras and the platforms dislocated by faulting. The same associations of forms are found in all regenerated geosynclines with, of course, variations in disposition resulting from their particular architectural and tectodynamic characteristics.

Relief of geosynclines in the residual phase

Progressive stabilisation, opposed by isostatic movements, engenders conditions which resemble in certain respects those of zones of deepseated folding in regenerated geosynclines. On the uplifted sections, erosion reaches deeper and deeper strata, made up largely of the metamorphosed material found at depth. However, some sediments persist, often caught up in downwarped strips. Some local planations cut into these massifs and, when the climatic and general morphogenetic conditions are favourable, they can reach an appreciable size. But they are quite soon affected by deformation, as unloading by erosion maintains a positive isostatic movement. The debris furnished by erosion accumulates in neighbouring basins which are then depressed, thus accentuating the warping of the margins. Regional tensions sometimes transform the warps into faults along the line of the main flexures. Volcanic outbursts are often produced in this way.

The main differences from the deepseated fold zones of regenerated geosynclines are as follows:

1. The reduced vigour of vertical movements, which accords with the general trend, namely one tending towards stabilisation. The downwarps

are less pronounced, and the faults less numerous and more frequently replaced by flexures and, above all, by warping.

2. The absence of a sediment cover of platform type overlapping the shield discordantly. The patches of sediments scattered on the massifs are the remains of geosynclinal sediments or of local detrital formations affected by uplift following their emplacement.

The Rocky Mountains in the west of the United States provide an example of the latter type of massif. The residual phase of the geosyncline commenced there about the beginning of the Eocene and has been prolonged for an abnormally long time because of their great extent from summit to base which, in favouring the accumulation of debris, maintains the compensatory isostatic movements. Moreover, within a framework of regional tension, certain tectonic paroxysms have occurred, even as recently as the Tertiary. They have somewhat regenerated the dormant geosyncline and have introduced some elements, albeit minor, of the same origin as those of regenerated geosynclines.

Erosion has been extremely intense, so that the massifs are formed of the crystalline basement, with sedimentary aureoles on their periphery. These aureoles are strongly folded and show subsequent faulting. They disappear in their turn beneath more recent detrital accumulations of the basins, which are very thick, the oldest being the most warped. The following series of types results:

(*a*) Crystalline or schistose massifs which retain no record of erosion surfaces older than the Cenozoic because, before this epoch, tectonism was too intense. However, they give rise to massive forms, often presenting accordant summits. But it is necessary to invoke the Davisian system wholeheartedly to see these as the remains of a Tertiary peneplain. In fact, these accordant summits, some lowered with respect to others, are the result of local planations of varying ages, developed in conditions momentarily favourable, and then dislocated by the continuation of tectonic movements. The essential difference between them and the deepseated fold zones of regenerated geosynclines is the absence of exhumed erosion surfaces by the stripping off of a discordant cover. Likewise, metamorphism has homogenised the lithology such that Appalachian forms do not develop in this material of the geosynclinal depths. The sedimentary fragments wedged in the centre of this zone have themselves become too indurated to be sculptured in this way.

(*b*) Border regions which are extremely complex because their beds have been affected by numerous successive tectonic paroxysms of varying styles. Folded, faulted, flexured, traversed by dykes and volcanic vents, they have sometimes been truncated by pediments which have been, in their turn, fossilised beneath discordant, detrital covers, transgressive outward from the basins. The oldest of these sedimentary covers have since been warped and sometimes faulted but never folded. They are uplifted on the margins

94

of the massifs and dissected in their turn, giving rise to homoclines: they disappear, truncated or not by new planations, beneath other, newer detrital series. Several successive detrital series can also terminate in a feather-edge on the margins of massifs which decline fairly gently. When the contact is sharper, these detrital series can be tilted at an angle of 20° along the flexures. In the detrital series made up of marls, sands, clays and conglomerates, homoclines of hogback type, etched into a herringbone pattern, characterise this type of structure.

Cordilleras, massifs of residual geosynclines and deeply folded zones of regenerated geosynclines make up, therefore, a morphostructural series which presents a certain unity arising from the essential role played by the tectonics of the structural block and from differences arising from an unequal rigidity. This series contrasts with that of the chains of ejective style.

Chains dominated by the ejective style: thrusts, thrust blocks, klippen

In the median mass the effects of compression, which are particularly intense, have often developed a tectonic style characterised by a dislocation of the beds: a mass of beds of variable magnitude becomes detached and is pushed well beyond its original position. This is the *ejective style*. Overthrust blocks and thrust splinters are its most frequent structural varieties. The klippen, well represented in the Polish Carpathians, are isolated masses of rocks, different from the rocks beneath and surrounding them: they are structural erratics which have been interpreted as the remains of deeply dissected overthrust blocks which do not always display their *roots* (that is to say, rock outcrops of the same structural units whose ejected part has been transformed into an overthrust block).

Of course there exists a transitional series from *rooted* structures, blending with the substratum, and the ejective style of thrust blocks. Wedge tectonics with recumbent blocks overthrusting the neighbouring structural units are a principal element. The same is true, but to a more marked degree, of thrust blocks, as may be seen in the Apennines. In this case, the thrust blocks are fragments of the calcareous or sandy substratum of Mesozoic age which have broken through the superjacent geosynclinal sediments and reached the surface where they form allogenic masses whose relief resembles, on a larger scale, that of the klippen.

COMMON CHARACTERISTICS

All these types of structures present two characteristics in common which have repercussions for the relief to which they give rise: tectonic vigour, and discontinuity of strata.

95

Tectonic vigour

Earth movements subject the rocks to considerable stresses, particularly when the deformations occur at depth beneath a heavy load of sediments. The rocks are fragmented, splintered, interbedded sometimes with sills and recrystallised. When the recrystallisation is fairly regular, which is rare, it indurates the rocks. But, more often, these stresses render the rocks heterogeneous, creating discontinuities within their mass which favour a differential fragmentation giving irregular relief, with aligned hills running along the limbs of folds. The cellular dolomites of the Alps are a particularly useful example. Fine rock cornices are rare in members having an ejective tectonic style. Cornices are generally replaced by aligned mountain stumps. On a smaller scale, at dimensions of about a kilometre, the beds of the same member are fragmented by secondary stresses, fractures, crushing and folding of great complexity. They accentuate yet again the heterogeneity of the rock, indurated by compression, further fragmented into breccia and stretched elsewhere and transformed into enormous boudins (*amygdales*) separated one from another. The planes of contact of the structural units present the same variety as do slickensides, though larger and more marked, the weight of the superjacent beds augmenting the friction when movement takes place on an inclined structural surface. The beds often become compressed, stretched and crushed. Some detached splinters interpose themselves due to friction between the two opposed blocks. They can reach a few metres to more than a kilometre in size. Tectonic breccias are frequent, but with a very varied geomorphological disposition. Badly cemented, they are more easily attacked than the normal rock. But, when well recemented, they can be extremely hard, like certain conglomerates. Thus cementation due to the circulation of subterranean waters often varies markedly over short distances. Here a breccia if hard gives rise to a precipitous rock face while elsewhere, where it is more easily broken down, it gives rise to a ravine or to a scree. The breccias which are formed at depth, under strong pressure, are more often indurated by a compressed cementation, while the breccias developed near the surface remain more or less uncemented. The same thing applies at the base of nappes and thrust splinters. Compressed and poorly recrystallised, the rocks which mark the trace of an irregular shear plane are readily fragmented and yield a series of screes while the beds situated at higher levels are less fragmented and form a rocky precipice. But there are other irregular contact planes, where a very strong recementation develops, which do not express themselves in the landscape. Such is the case in the crystalline nappe terrains of the eastern Alps.

Accordingly, tectonic vigour is a source of heterogeneity in the landscape. It proliferates differences over short distances. The disposition of the rocks is no longer solely a function of their initial lithological characteristics but, for the most part, of modifications impressed upon them by tectonics. Now these modifications vary over slight distances as a function of the mechanical conditions prevailing from time to time during the deformations. The differences from bed to bed are rendered indistinct by these less constant and differentially disposed variations. This has repercussions which directly affect the landscape, varied in its details at the scale of single folds, but where the major lines of tectonic deformation are frequently effaced by local structures and appear less clearly than in the folded cover rocks.

Discontinuity of strata

Subjected to considerable stresses, the beds lose their initial continuity. Some of the more rigid beds are fragmented and crushed, facilitating their deformation. Other more plastic materials, such as the clays and the thinly bedded flysch members, are stretched here and thickened there. Compaction effects are produced such that beds are folded over again upon themselves. Further, numerous ruptures traverse each group of beds. If to that is added the modifications arising from tectonic stresses, collapses and recrystallisations, we can appreciate the complexity of detail which results. In this type of structure it is exceptional for two superposed beds to maintain their initial parallelism. Each has responded in its own way under the effect of tectonic stresses. There is disharmony.

Structural disharmony, very general and particularly well developed in areas of ejective style, has important geomorphological consequences. In the course of their incision into uplifted blocks streams cut into a structure which does not remain the same throughout the total thickness of the beds traversed but, on the contrary, is constantly changing. There is, therefore, a permanent superimposition of drainage. The general form is controlled by orogenesis and causes the streams to flow from the most elevated points towards the lowest points. But, in detail, forever running across different architectures during incision, the stream courses show a succession of local adaptations and discordancies which may be described as chaotic.

Discontinuity of beds does not exist only at the detailed scale, some elements occurring in each architectural unit. Discontinuity occurs also on a global scale between diverse structural units, e.g. between several overthrust blocks piled up one on another such as is frequent in the Alps, or between a nappe and its substratum or between a fault block or a klippe and the root zones around them. These discontinuities between architectural units are expressed more or less clearly in the relief. They appear clearly in the case of klippen and fault blocks because in this case they consist of masses of resistant, coherent rocks rising in the midst of plastic formations of mudstone shale type. In contrast, they are completely undetectable in the relief in the crystalline and metamorphic nappes of the eastern Alps (Pennide nappes) and their interest, then, is exclusively geological. But in the present case it is a question of deep structures having been brought to

97

the surface as outcrops at the end of a long period of evolution: it is at the beginning of the residual geosyncline stage. In the younger, eastern Alps, the nappes are reflected in the relief. The correspondence between architecture and relief is modified, therefore, in the course of geosynclinal development. Correspondences are clearer at the beginning of the orogenic phase and become attenuated subsequently in the course of stabilisation, due to the outcropping of deeper and deeper structures whose beds have lost their lithological individuality due to crushing, recrystallisation and metamorphism. Following that period of geosynclinal development during which they are brought into outcrop and dissected, similar structures can yield quite a different relief type.

These common characteristics, bound up as they are with the development of the geosynclinal assemblage, are more or less clearly visible depending on the type of structure. Let us examine these now.

THE RELIEF OF OVERTHRUST BLOCKS

Overthrust blocks are a structural entity. They do not correspond to a geomorphological type. In fact, they give rise to a whole series of different relief forms. The deep, metamorphosed nappes are not expressed in the relief in their original form. They are practically homogeneous. Their rigid

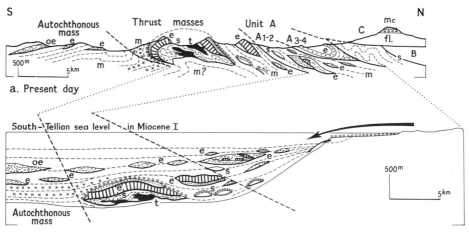

Fig. 2.8. Nappes in the southern Tell: schematic sections by A. Caire, 1962

The top section shows the present condition, with nappes and *écailles* (splinters). Some of the latter are eroded as klippen (Eocene thrust masses to the south of 'Unit A').

The bottom section is a hypothetical reconstruction of the state of affairs at the beginning of Miocene I, with a geosynclinal furrow in which slipped masses from the north are accumulating. The Miocene geosynclinal furrow has been squeezed and covered by the nappe of Unit A. It has been folded, with klippen moving in relation to the enclosing sediments.

The letters are the initials of the geological stages (t. Trias; s. Senonian; e. Eocene; o. Oligocene; f. Flysch; m. Miocene; mc. Miocene conglomerate) material is reworked by *plis de fond* and gives rise to wedges which evolve in the usual way. Hence, there is little difference between the structural relief of the Central Austrian Alps and that of the crystalline High Atlas or certain Andean wedges. At the other end of the series, the sedimentary nappes, made up of sedimentary cover rocks of the platforms and showing little dislocation (such as the Helvetic nappes of the Swiss Prealps), present a relief of culminations and homoclinal dips strongly resembling those of the autochthonous French Prealps (Genevois, Bauges and Chartreuse). In their alternations of limestones and marls, the same relief forms are found. This is because the beds are of the same nature and the tectonism is of the same style. In both cases it is a matter of the covering rocks having become detached from the substratum (*décollement*). The significance of such detachment varies, but this problem is of a geological rather than a geomorphological nature. The essential point is that it has been accomplished sufficiently gently for the same style of folding to be found, in both cases, in similar material.

The intermediate case is more exceptional: varied sedimentary rocks are detached and ejected into superficial nappes but in this case the dislocation is sufficient to minimise the similarities to the cover rock folds. We give an example below.

From the Chablais to the Dachstein, the northern flank of the Alps is bordered by overthrust blocks made up of flysch, marls and limestones. Rigidity varies considerably with changes in the proportion of the diverse facies. In Austria, for example, the nappes are largely calcareous and rigid. In the Swiss Prealps, on the contrary, some sections show a convergence of forms with the folds of autochthonous cover rocks. Accordingly, it is necessary to study relief forms resulting from overthrust blocks as a function of differences in tectonic style and evolution.

The first example, that of rigid nappes, is provided by the calcareous massifs of the Austrian Prealps. Feebly warped Triassic beds, reaching a thousand metres in thickness, have suffered tangential pressures with scarcely any folding. They were moderately tilted, toward the outer part of the chain following block uplift of the axial zone. The fractures dominate the landscape, but have now largely been levelled by a Miocene erosion surface (*Raxlandschaft*), sprinkled with crystalline pebbles derived from the axial zone and finally deformed by the Pliocene–Quaternary orogenesis. The general form of these nappes is that of calcareous plateaus affected by fractures which isolate secondary blocks showing advanced karstic forms, and whose summits constitute part of the erosion surface of the Rax, degraded and tilted towards the foreland. These calcareous plateaus make up the backslope of a substantial homocline which faces the central massifs. The front of this great structure is made up of the calcareous nappe and by the subjacent plastic formations which have allowed its *décollement*, as shown by a peripheral groove. The disposition is the same as that of the Alpine furrow.

FIG. 2.9. Folded strata in a nappe, in the Diablerets massif, Switzerland
Jurassic limestones of the Helvetic nappe, folded into a buckled anticline.
The axis of the anticline is at the summit on the right side of the picture. The
limestone bed is much fractured, so that the cliff that it forms has not the
same rigidity and constancy as the cliff-scarps of the autochthonous strata

In the nappes which are sufficiently rigid not to have been too severely
overfolded during their movement, the homoclines are clear and well
developed. They are encountered in great numbers in the Helvetic nappes,
where they are always calcareous.

A second example is provided by the strongly refolded nappes such as
those of the Briançonnais. Like the pre-Alpine nappes of Austria and
Switzerland, these nappes are in part made up of limestones which alternate

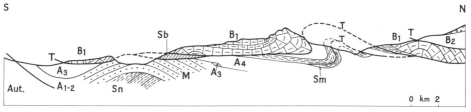

FIG. 2.10. Schematic section across the Tellian nappes (on the Bordj-bou–Arreridj
map-sheet (*after Caire, 1962*)
On top of the autochthonous strata, which outcrop on the southern margin, are piled
two nappes. Each of these nappes is subdivided into several plates (A1, A2, A3, A4 and
B1, B2). The Miocene conglomerates rest uncomfortably on the Senonian (Sn), and are
themselves overlain by plates A3 and A4. The relief is due to limestone masses that
'float' on beds of marl.

with more supple facies of flysch and *schistes lustrés*. The obstacle to their movement offered by the Hercynian crystalline massifs forced them to become vigorously overfolded. They are strongly overfolded on themselves, in accordion fashion. The anticlinal axes have generally yielded to tensional stresses and have, in this way, been more readily eroded. In the synclinal parts of the overfolds, on the other hand, the beds have been subjected to compression which in general has indurated them. Thus, they offer more resistance. But being strongly disturbed and fragmented, these nappes do not give rise to good structural relief. The clearest forms are summits or massifs which correspond to the synclinal parts of the overfolds as a result of inversion of relief. This Briançonnais type, with summits made up of synclinal overfolds, yields a relief more confused than the nappe homoclines. Generally speaking, in the nappes the synclines are more resistant than the anticlines; the beds are less fragmented there than in the anticlinal axes where they habitually collapse. In the Helvetic nappes the large synclines, made up of relatively regular and undisturbed beds, are well preserved and terminate subaerially in crests which form the highest summits: these dominate the anticlinal axes where the beds, vigorously tilted and fragmented and violently eroded, give rise to a confused and accidented relief with small discontinuous crests and isoclinal ridges.

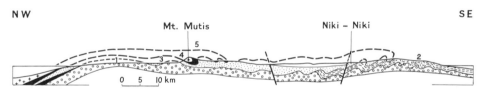

FIG. 2.11. A geosynclinal chain formed by the piling up of nappes (the 'hyperplastic' form) in western Timor (*after Lemoine, 1958, p. 209*)

The numbers represent several nappes. The beds have been so severed that they have lost all contrasts of hardness, and this complex structure does not produce great relief, apart from the homoclinal ridge of Mont Mutis, formed by a klippe of crystalline rock belonging to nappe 4, wrapped round by nappe 3.

Finally, in certain cases nappes may give rise to no distinctive relief forms even though the geosynclinal development be not too far advanced. In the Rif, for example, where some structural geologists continue to multiply the nappes of distant, Spanish origin, nothing of this is expressed in the relief. In effect, these nappes are formed of shales insensitive to differential erosion, the relief being controlled by tectonics, by the structural undulations and the manner in which they guide a particularly violent dissection.

The variety of the structural relief arising from overthrust blocks is exceptional in the Alps. It is a result of the tectonic peculiarities of the chain: extreme compression and incorporation during the orogenesis of a platform margin made up of the Hercynian massifs and their cover rocks with a well differentiated lithology.

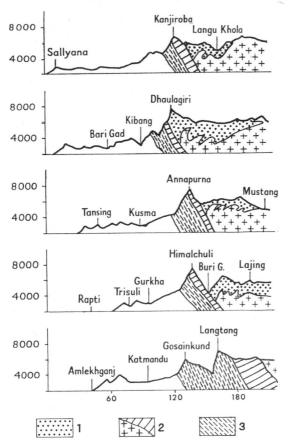

FIG. 2.12. A series of sections across the Nepalese
Himalaya (*after Hagen, 1954, p. 327*)
   Vertical exaggeration 5 times. The Tibetan
nappes are folded against the roots of the
Katmandu nappe, which stands up boldly giving
homoclinal ridges
1. Sedimentary rocks of Tibetan zone
2. Katmandu nappe and crystalline basement of
   Tibetan zone
3. Roots of Katmandu nappes

THE RELIEF OF KLIPPEN AND FAULT BLOCKS

The klippen and fault blocks present, by definition, a sharp lithological
contrast with the neighbouring structures: this contrast is always to some
degree expressed in the relief, typically in a striking fashion.

The klippen

These are masses of resistant rocks, usually calcareous, coherent and little
fragmented, which give rise to a rocky relief with sharp crests or little

Fig. 2.13. The front of a nappe, near Sallanches in Savoy
  The limestones of the nappe, outlined by the snow cover, rest on slates,
which have soft surface forms and cultivable soils. Only gently inclined, the
nappe forms a homoclinal ridge

plateaus bounded by escarpments, resting on rocks of only moderate
resistance such as shales, flysch or marly series.

  The klippen do not have roots. Beneath the coherent rocks of which they
are formed, the more plastic formations of the vicinity extend, providing a
violent lithological contrast and allowing the klippen to persist in the form
of residual relief forms rather in the manner of outliers. The finest examples,
formed of limestones, surmount the flysch of the Polish Carpathians.

  Klippen, forms reaching some kilometres in size, seem to have several
origins. Some workers see them as the remains of glissading strata, notably
collapse structures, coinciding with the rock masses which are least dislocated and have been able to offer a certain resistance to erosion. Others
see in klippen the remains of overthrust blocks entirely dismantled by
erosion, but difficulties arise here when attempts are made to trace the
roots of these nappes! It seems that klippen are often formed by lenticular
facies, e.g. old coral reefs buried in clay–marl beds which, under the effects
of earth movements, have behaved differentially thus accentuating their
distinctiveness.

The fault blocks

These have been particularly well studied in the Apennines, where the type
has been defined. In this region is found a Mesozoic substratum made up of
limestones and marls emplaced under conditions of platform sedimentation

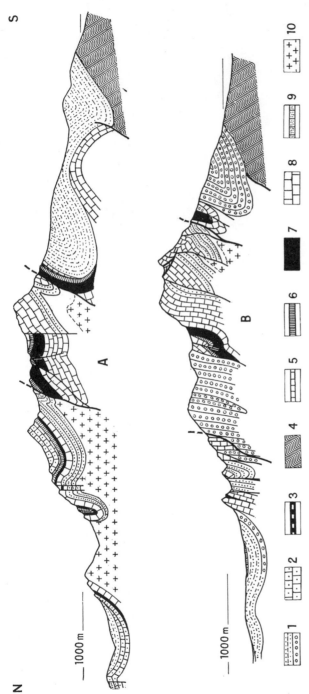

FIG. 2.14. Sections across the Djurdjura, showing fractured recumbent folds *(after Flandrin, 1952)*

In A, a series of plates in the central part, with masses of hard limestone giving strong homoclinal ridges, facing in the direction of the thrusting movement. In the north, a small asymmetric syncline is broken on its upper limb. In the south, overfolding of the Djurdjura mass on its Oligocene border—unconformable beds of conglomerate, marl and sandstone formed by the disintegration of the rising mountain chain. The Oligocene beds, only affected by recent tectonic activity, are much less deformed than the Mesozoic strata. The folds are more open and the rocks more pliable, but in the absence of marked lithological contrasts the folds produce no relief.

In B, the compression has been more violent. The Mesozoic strata are sliced into large flakes, dipping almost vertically. The Oligocene conglomerates are strongly folded, sometimes also fractured and tilted vertically. The relief is a general upwarp, with ridges corresponding to the harder beds. The detail of the folds is not much reflected in the relief, which is produced by the uplift as a whole.

1. Oligocene
2. Lutetian
3. Albian
4. Cretaceous flysch
5. Berriasian
6. Tithonian
7. Upper Lias
8. Lower Lias
9. Pre-liassic
10. Crystalline schists

followed by Tertiary geosynclinal development during which sedimentation took place, filling the grooves with marls, flysch and more or less sandy clays totalling several thousands of metres in thickness. There followed a tectonic paroxysm which, in the Miocene, transformed these grooves into cordilleras such that new grooves formed by downwarping on their margins. The relatively rigid substratum became dislocated. Certain sections were thrust upwards, more or less obliquely, piercing the soft infill.

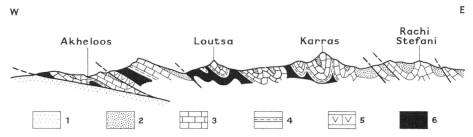

Fig. 2.15. Recumbent folds ('ejected' style) in the southern Pindus, Greece (*after Antonin, 1959*)
   A series of overthrusts with irregular contacts. To the west, they produce homoclines. The remainder of the section is strongly folded and shows essentially the bases of the synclines: the beds have been forced up by compression between the synclines to produce crushed anticlines which, for that reason, are readily destroyed. Two of these synclines stand above the other mountains and form perched synclines, but the inversion of relief is only theoretical because, given the style of folding, the anticlines have *never* constituted the dominant relief elements. The low proportion of weak rocks and the recency of the uplift explain the rather limited degree to which the synclines have been etched out.

   The fault block is also characterised by an important vertical tectonic component, which is usually predominant. In a sense, this is a horst. But they always suffer, as well, a certain tangential stress, which tilts them and causes them to overthrust on one side the formations across which they have moved. Rigid, lenticular formations can be subjected to the same process: fault blocks pass into klippen.
   By definition, the fault block is formed by rigid strata having moved *en bloc*. It is particularly affected by faults which sometimes subdivide it into unequally uplifted masses. These are *composite fault blocks*.
   The fault block constitutes an isolated relief form, giving rise to ridges or sharp edges of readily dissected country rock surrounded by rocky escarpments. Little fractured, it results in a plateau which is generally triangular and tilted. Larger and subdivided by a series of fractures, the composite fault block is more complex. The Gran Sasso of Italy, for

example, combines large fault escarpments with unequally uplifted and diversely tilted fault splinters.

It is known also that fault blocks have another origin, as shown by Flandrin in the Djurdjura of Algeria. A platform margin was at first faulted and then affected by tangential movement. This, more intense at depth, thrust up the horsts in compressing them at the base. They were transformed into fault blocks and thrust splinters. The main difference between this origin and the standard explanation of fault blocks lies, perhaps, in the absence of a superficial cover through which the ejection is made.

Like klippen, fault blocks have dimensions measurable in kilometres. In effect, since they are larger, they break up and become composite fault blocks. These reach about 10 km in size. It is a question, therefore, of smaller relief forms than in the case of most overthrust blocks which usually exceed 10 km in breadth.

Having studied several types of massif, it falls to us now to say something about subsidising sectors, the intermontane basins.

The intermontane basins

The intermontane basins correspond to tectonic elements which have subsided and which constitute also regional base levels. Accumulation is their prime characteristic: but their tectonic character is continually changing and the basins are often warped resulting in a dissection of their infill. Therefore, we can see successively the phenomena of accumulation followed by dissection.

ACCUMULATION IN INTERMONTANE BASINS

Subsidence reduces the gradient of water courses and stimulates at least partial deposition of alluvia which builds debris cones (*cônes de déjection*) at the foot of mountains. If the basin is sufficiently large, the finer materials, sandy silts or even clays are deposited in the distal areas. Under dry climates, evaporation can cause the precipitation of dissolved matter giving *evaporites* (salt, gypsum, potash, etc.).

In contrast to piedmonts, intermontane basins are bordered by mountains on all sides. In order to escape, the rivers must traverse an uplifted block so that they are generally incised in gorges. Their downcutting is slower and more difficult the more rapidly the block is uplifted and the more massive and coherent the rocks. The intermontane basin, therefore, is reduced by a flattening of the river's long profile between sectors with steep slopes in the headwaters and in the lower reaches. The flattened sector may be practically horizontal and even become replaced by a basin, if the block situated at the downstream end is uplifted more quickly than the river cuts down. In this case, a lake or a marshy area occupies all or part of the intermontane basin. The alluvial cones are extended by the spreading of alluvium in the central part of the marshy area of the depression or by subaquatic deltas in the lake. In dry climates such lakes may be without outlet, evaporation removing the excess water and preventing the lakes from overflowing by way of gaps in their bordering mountains. The marshy zones may be periodically inundated by floods and then dried by evaporation. It is in these conditions that evaporites are deposited, generally alternating with silty beds which settle out just after the flood before the level is progressively lowered.

The intermontane basins also play the role of a sediment trap favouring a thick infill which is generally several hundreds of metres thick and sometimes more. The nature of this infill varies according to the climate: the evaporites demand a dry climate more easily achieved thanks to the mountainous screen which surrounds the basins. But vigorous erosion, under all climates, guarantees an abundant supply of detrital materials. The proportion of pebbles, sands and clays varies as a function of climate, the steepness of slopes and the gradients of streams, and the nature of the rocks. It changes, therefore, in a clearly defined basin, under the influence of climatic oscillations which have had special importance in the Quaternary. Volcanic eruptions, in destroying the vegetation and providing fresh material, particularly emissions of cinder and scoria, often stimulate massive transport of alluvial material. Reworked by water, their products form hydrovolcanic accumulations. They are frequent in the intermontane basins, more especially as these often owe their origin to dislocations which, in other respects, favour the outpouring of magma.

Sedimentation in the intermontane basins is therefore extremely varied and influenced by:

1. the tectonic evolution both of the basin and of the block situated downstream which controls the inflow and the outflow, and therefore the possibility of the development of lakes, marshes or simply alluvial plains. The relative earth movements also control the total thickness of accumulations;

2. the existing climate in the mountains forming the basin slopes: this influences the nature of the material transported and the mode of sediment transport e.g. fluviatile or glacial;

3. the methods by which the mountains forming the basin margins are dissected: these are influenced by tectonic movements, by the nature of the rocks and by the climate.

In most cases the detrital loads of streams are sufficiently abundant to yield, on the margins of the basin, a zone of alluvial cones. When they are sufficiently developed, they overlap one another. Their slopes are relatively steep. Their volume depends upon the vigour of the dissection of the neighbouring mountains. These cones are larger at the foot of the most severely dissected block, which is usually the most uplifted. Very often, the filling up of the basin is therefore asymmetrical, with large alluvial cones spread broadly at the foot of the highest mountain and many small cones at the foot of less uplifted blocks. Longitudinal streams in the basin are often pushed over against these smaller cones. This is the case of the Mapocho in the basin of Santiago de Chile.

When the basin is sufficiently large, the cones give way progressively downstream to finer and finer sediments on gentler slopes. The centre of the basin is then occupied by a very uniform swampy plain. The general appearance of the sedimentary fill is concave, and may be asymmetrical.

But the basins and the neighbouring blocks are subjected to tectonic movements which affect the fill as much as does the drainage downstream.

Tectonic deformation and dissection

Tectonic deformation does not act alone. It is always associated with the factors controlling surface dynamic development. This results in a great variety of intermontane basins. Rather than attempting to define every type, we shall stress the processes necessary to their understanding.

Deformation and infilling

The intermontane basins are always located along lines of structural weakness, habitually on fault zones which develop in areas of tensile stress. Also, they are subdivided into compartments by faults. Their depression is unequal. Some fault directions are such that these compartments lie obliquely to the general direction of the trench. Thus in the Santiago Basin, orientated north–south, a secondary horst is aligned southwest–northeast and subdivides the basin. During part of the Quaternary, a lacustrine basin was isolated in the extreme north of the basin, for no important stream debouched there. In the south, on the other hand, the infilling is made up of fluvioglacial cones and deposits of glacio-volcanic outbursts. But the secondary horst was eventually engulfed by the sedimentary infill and the Mapocho River, in passing over it, then went on to build its cone in the northern lacustrine basin.

These secondary block movements in the intermontane basins play an important role in the emplacement of the deposits. As has been shown by researchers in Hungary (the largest of the intermontane basins of Europe), water courses move towards the most depressed zones, and this favours the accumulation of alluvia and, up to a point, some compensation of subsidence by infilling. In effect, on the surface of the fluvial deposits, cones or paludal areas, streams shift their courses readily and a tectonic stress, even when unobtrusive, can provoke a diversion during flood and the subsequent maintenance of a new stream course.

Deformation and evolution of basin margins

The margins of intermontane basins are critical zones, most often controlled by faults and sometimes by flexures when the substratum is sufficiently flexible and the earth movements not too sharp. The tectonic activity there is generally intense while, on the other hand, the severe break of slope stimulates accumulation which takes the form of debris cones (*cônes de déjection*).

There are two principal cases. First, that in which the tectonic movements are intense and the basin-block mountain differentiation tends to be accentuated. Faults and flexures are frequently reactivated. The cones are tilted, the more so the older they are. The longitudinal profiles of the cones are cut again: the upper part of an older, uplifted cone appears rapidly above a younger cone and on its upstream (mountain) side, while the lower part, which has been depressed, plunges towards the basin beneath this same younger cone. At the foot of the Andean cordillera, the eastern margin of the Santiago Basin displays some magnificent examples. But it often happens, also, that the bordering zone of flexuring is displaced and encroaches either on the mountainous zone or on the basin. When it encroaches on the mountains, the zone of subsidence extends and the old sedimentary accumulations disappear in the depths. At the surface, it can be seen that the recent cones reach the mountain front without the interposition of ancient cones. If, on the other hand, the zone of subsidence becomes restricted in extent, the old accumulations are uplifted along the border of the mountains with which they become incorporated and a fringe of uplifted, dissected old cones appears, in various stages of warping. Moreover, their headward parts, rapidly denuded, are more clearcut towards the upstream end than those of the recent cones thrown out downstream, in the zone which has not yet been affected by uplift.

The tectonic movements are slow and denudation takes the upper hand, given favourable lithological and morphoclimatic conditions. Under climates favouring general subaerial modification as, for example, dry climates with diffuse stream channels, local planation surfaces can then develop on the margins of uplifted blocks. Most of the pediments of the western United States developed under such conditions. The persistence of a slow subsidence of the basin favours their development since it permits a progressive evacuation of material as it is liberated by disintegration and prevents any great thickness of accumulation on the sediment-veneered slopes of great uniformity (*glacis*). Such local planations do not demand tectonic stability, only a certain dynamic equilibrium between a moderate rate of subsidence and the activity of a particular morphoclimatic system.

Deformation and basin drainage

Very often, intermontane basins have floors well above sea level. The one at Santiago de Chile is at about 500 m above sea level, that of the Cerdagne at 1 100 m, the Caracas basin at 700 m, and the one at Lake Titicaca at 3 900 m. They continue to act as sediment traps only for so long as their outlet stream does not succeed in cutting down through the uplifted block which encloses the downstream end of the basin. Schematically it comes down to a problem of relative rapidity of process. On one side, there is the speed of linear incision of the talweg of the outlet stream: on the other, the rate of uplift of the tectonic barrier. Now both vary considerably according to the situation, and in the same situation as a function of time. Tectodynamism is typically staccato. Climatic oscillations, on the other hand, have repercussions on the rate of stream incision. Periods when rapid modification of slopes produces layers of debris of excessive number or coarseness may halt the incision of talwegs. However, the incision increases again when the supply of debris is reduced and when the calibre of the debris allows their ready removal. Certain climatic oscillations check stream erosion and give rise to climatic terraces while others permit stream incision to continue. The dry climates, which are marked by considerable evaporation in the intermontane basins, are unfavourable to this incision of outlet channels. Such climates can even prevent the formation of an outlet stream, the basin then being occupied by sheetwashes or, if the water is more abundant, by a lake with oscillating levels. Many basins in the western United States, now transformed into saline depressions (*playas, schotts, sebkhas*) were occupied by lakes during the pluvial episodes of the Quaternary. Under these conditions, sedimentation persisted for a considerable time during the course of active tectonic development. A stream had to develop and incise the margin of the basin by retrogressive erosion in order to drain it, as was the case with the Chélif River in Algeria.

Under more humid climates, lakes persist only during the initial period of diastrophism. Thereafter, their outlet stream cuts a gorge through the tectonic barrage and they are drained off. This is the case with the Pannonian Basin. For some fifteen million years it was the site of a vast lake in the Sarmatian stage of the Miocene, then the Danube was formed and, cutting the gorges of the Iron Gate, it drained the lake and instigated the dissection at the downstream end of the basin's sedimentary infill. The general tendency to erosion, interrupted by phases of climatically induced deposition, has produced, around Belgrade and downstream of it, a series of paired terraces. In the centre of the basin, which is still being downwarped, the Quaternary saw the accumulation of almost a thousand metres of sediments, their base lying well below sea level. This fill is not dissected, for the general subsidence has, in effect, eliminated all retrogressive erosion. Finally, in the vicinity of Budapest, warping on the edge of a horst has given rise to truncated and intersecting debris cones.

The evolution of intermontane basins, therefore, makes clear the necessity for inductive reasoning in assessing the opposed effects of tectodynamism and morphogenesis, itself influenced by tectodynamics as well as by climatic oscillations which are partially independent of the latter. Each concrete example demands a reconstruction of the interplay of these diverse factors and their variations, based on the study of the sediments and on detailed mapping of their disposition.

The intermontane basins, being bound by structural elements, form during various phases of geosynclinal development. The geosynclinal furrows can produce them, when they do not open directly on to the sea. This is also the case in the orogenic phase, for uplift of ridges in the geosyncline is always accompanied by localised adjacent collapse, especially in periods of release of tangential pressures. Finally, progressive stabilisation of the geosyncline, in multiplying block movement, is particularly favourable to their initiation. Regenerated geosynclines are characterised by well-developed intermontane basins. The study of their sedimentary fills is one of the main sources of information used in the reconstruction of the development of mountain chains. Moreover, this is essential to the understanding of their structure.

General structure of the geosynclinal folded chains

The complexity of geosynclinal development results in a great variety of structural characteristics in each constituent chain of the geosyncline. Nevertheless, some general problems of structure, resulting from the processes just discussed, are discernible. It remains to examine these, it being clear that we must limit ourselves to the evidence most likely to assist an understanding of the true origin of each chain or part chain. We examine in turn the disposition of architectural elements and the hydrographic network which controls the mode of regional dissection.

ARRANGEMENT OF THE STRUCTURAL UNITS

Here again, the difference between intracontinental and littoral (unilateral) geosynclines is evident.

Continental geosynclines

The effects of compression are important because of the existence of rigid barriers in the form of continental masses. With slight displacements of these units (not to be confused with continental drift) compression effects, constricting the continents and folding the strata, alternate with tensional effects which are accompanied by tear-faulting and subsidence. The disposition of the structural elements is influenced by these particular types of development.

During certain paroxysmal disturbances, the geosynclinal structure advances on the neighbouring platforms. These become fragmented and their debris is then incorporated in the chain. This is true of the Hercynian massifs of the Alps, and of their cover of platform sediments preserved in overthrust blocks. Geosynclinal instability can reactivate ancient, largely stabilised geosynclinal structures and give rise to regenerated geosynclines. The Rockies, which are included in this type, are in fact an intracontinental chain if one takes account of the existence on their western flanks of the rigid element of the Colorado Plateau, a pre-Cambrian shield surfaced by tabular platform sediments whose tectonic role resembles that of the Tibetan Plateau.

On its margin, tangential movements of the geosyncline can affect the sedimentary cover of the platforms, yielding folded cover rocks if they possess near their base a formation sufficiently plastic to allow *décollement* to occur. This is the situation in the Jura, where a tilted 'keystone' wedge occurs at the contact between the Jura and the Alpine geosyncline.

These phenomena create structural relationships clearly seen in the Alps: long-studied structures which are an established type. Geologists have succeeded in distinguishing:

1. An axial zone, made up of an ancient geosynclinal furrow and fragments of the Hercynian basement. The infill of the furrow is of Cretaceous and Eocene *schiste lustré* and flysch sediments, severely folded and overthrust. The Hercynian fragments give rise to the granitic and metamorphic central massifs. Within the ensemble, the latter are more resistant than the geosynclinal sediments and make up the more elevated massifs. But in all these crushed rocks, the architectural forms are unclear and little developed. The relief is confused, with great valleys as the principal characteristic, controlling the pattern of dissection.

A transition region is that which runs through Gap (Lower Maurienne), an ancient platform margin on which somewhat more varied sedimentation has occurred, although it remains rather monotonous with a substantial series of Mesozoic blackish marly limestones, and black marls ('Dauphinois facies'). In the Embrunais, for example, where the central massifs have not blocked their advance, the area is covered by overthrust blocks. Elsewhere, it has been revealed by differential erosion, yielding the Alpine Corridor (*Sillon Alpin*).

2. An outer zone, formed by the platform margin subjected to tangential forces damped down by the transition zone which acted as a buffer. This zone is the Pre-Alps (in the geographical sense). Up to the middle Cretaceous it was marked by the deposition of alternating limestones and marls resembling those of the Paris Basin. Thereafter, sedimentation became episodic and interrupted by discordancies, all repercussions of movements in the geosyncline. Moderate folding accompanied by faulting occurred in the Eogene and these structures were more or less bevelled and covered again with terrestrial detrital deposits. With orogenesis in the Neogene, the outer zone was overtaken by the molasse sedimentation which thickened towards the outer part of the chain. It was then uplifted and tilted outward, in its turn, and folded anew. It is at this time that gravity tectonics would have come into play, following such uplift.

3. A piedmont region, which occupied the site of the present Pre-Alps, and which was subsequently reduced in extent by their uplift. It received the debris produced by the dissection of the orogen, yielding the molasse sediments. It was downwarped marginally in the form of a furrow, under

the probable dual influence of displacements of material at depth and isostatic phenomena, permitting the laying down of a thickness of 2 000 m of deposits none of which are of deepwater facies. Then it was warped and contracted all at once during the uplift of the Pre-Alps. The part which remained depressed forms the series of basins making up the Rhodanian Corridor (*Sillon Rhodanien*).

However, all the intracontinental chains are far from presenting an entirely clear picture. Much depends on the regional tectonic framework and the precise tectonic mechanisms. Thus, in the Pyrenees, there is a very well developed axial zone which, east of Toulouse, is a regenerated geosyncline. Then follows a narrow band of compressed Mesozoic rocks, realigned Palaeozoic massifs with discordant cover rocks and, finally, the folded Mesozoic cover which plunges steeply beneath the Tertiary rocks on the edge of the Basin of Aquitaine. Vertical movements in the regenerated geosyncline are accompanied by compaction effects hardly conducive to the development of a typical outer zone. In the Tell of Oran massifs made up of the marly fills of geosynclinal furrows, severely folded and constituting a tectonic relief, alternate with subsidence basins where a substantial detrital series, often very coarse, is accumulating. To the south, the High Plateaus represent a fragment of the rigid shield with a thin sedimentary cover more faulted than folded. It is only beyond, in the Saharan Atlas, that the sedimentary cover thickens and is made up of plastic strata capable of suffering *décollement* and producing the characteristic folding. A similar distinction between chains with folded epicontinental covers, and massifs corresponding to the geosynclinal furrows separated by fragments of rigid shield, is common in the Mediterranean from Spain through Anatolia as far as Iran.

Littoral (unilateral) geosynclines

At the junction of the continental platform and the ocean basins, geosynclines are not affected by compression as intensely as the intracontinental geosynclines. The tectonic style is distinctly different. Overthrusts are practically unknown. The tangential movements produce at most only thrust splinters which develop in special conditions. Thus, in the Peruvian Andes, thrust splintering has affected a Mesozoic cover of limestones and marly limestones overlain by a substantial detrital series with volcanic, argillaceous, conglomeratic (*Capas Rojas*) and sandstone intercalations which cover a fragment of shield incorporated in the chain and carried up to an altitude of 3 000–4 000 m. On the Pacific Coast, it is bordered by the granitic batholith emplaced in the Mesozoic and covered by volcanic outpourings of andesites, and subsequently up-arched in a geanticline in the upper Tertiary.

The folded cover rocks are of little extent and generally not characteristic. They can be found locally in the Mesozoic sandstones and shales of the Venezuelan Andes, where they occupy moderately uplifted sections compressed between more uplifted wedges. But these folded sections are discontinuous, very broken and occupy only weak areas. They are a simple result of predominantly vertical block movement. In northwest Argentina the eastern flank of the Andes is composed of a band of Palaeozoic folded rocks. It borders the chain between the arch of the High Cordillera crowned by volcanoes, and the Patagonian–Pampas shield.

Accordingly, it is the vertical movements which are dominant. Littoral chains are thus made up of an assemblage of unequally uplifted blocks, separated by great fractures lined with volcanoes. In central Chile there is a Mesozoic granitic batholith reaching to a height of approximately 2 000 m (the Cordillera de la Costa); then a subsidence trench broken up into a series of basins by secondary horsts often tangentially disposed; then the High Cordillera whose central wedges attain an altitude of 4 000 m with very varied rocks (sediments, intrusives, old volcanic material) crowned by volcanic piles reaching to 6 000 m. It frequently happens that fragments of shield are incorporated in these volcanic masses. This is the case in the Lake Titicaca region where Palaeozoic shales, folded then overlain by Mesozoic limestones and grits, have since been buried in the Miocene geosyncline (4 000–5 000 m of sandstone–shale facies) before being raised to an altitude of 4 000 m and flexured. Some folding of cover rocks occurred. This is poorly developed in the Miocene rocks but is more evident in the Mesozoics where thrust splinters become apparent.

Similar relationships are met with in the western United States. On the edge of the Pacific the coast ranges are formed of a vigorously folded geosynclinal fill, with major dislocations. Then come the blocks, some collapsed like the Great Valley of California and others uplifted, like the tilted, crystalline mass of the Sierra Nevada. The Colorado Plateau is a fragment of shield with cover rocks raised to an altitude of 3 000 m and passing toward the south into the patchwork of fault blocks in Nevada. The Rockies are a regenerated geosyncline in which uplifted massifs and depressed basins alternate. They terminate overlooking the vast, warped piedmont of the Great Plains which have existed since the Cretaceous. A migration of folds towards the outer part of the continental block can be clearly seen in this case.

Finally, one other characteristic, vulcanism, is important in littoral chains throughout every phase of their evolution. In Chile the Jurassic is largely represented by volcanic material. In Java and Sumatra the chain at present in formation (bordering, on the edge of the Indian Ocean, the Sunda platform) is above all a volcanic chain. An outward migration of eruptive centres towards the Indian Ocean is notable. The degree of subaerial destruction of the volcanoes varies accordingly.

ORGANISATION OF THE DRAINAGE NETWORKS

The hydrographic net plays a principal role in morphogenesis. Deep valleys are often the most characteristic feature of the relief of geosynclinal chains.

The vigour of dissection depends on the stream network. It is this which dictates the contrast between the high plains of Lake Titicaca or of Algeria, and the wild gorges dominated by the maze of razorback ridges and long valleyside slopes of the massifs of Tell or the descent to the Pacific or the Amazon.

The relationships between the stream network, dissection and tectonic deformation are at the heart of the Davisian notion of the cycle of erosion. What we have seen of the genesis of geosynclines reveals how the Davisian hypothesis falls down: mountains are not raised initially, followed by the subsequent development of the stream network and increasing dissection such as can be seen on the muds of empty reservoirs: rather, there is permanent interaction between tectogenesis and morphogenesis. The basic conceptions being false, the terms created to express them become obsolete.

To express the relationships between tectogenesis and dissection as they saw them, W. M. Davis and his followers set up two terms: *superimposition* and *antecedence*.

(*a*) *Superimposition*. The stream becomes established on a particular structure and then, in the course of downcutting, encounters different structures. Its trace is etched on to the superficial structures resulting from tectogenesis prior to dissection. Being adapted to these, the stream is discordant to the entirely different structures at depth. Accordingly, superimposition stimulates a lack of adaptation which is expressed in the landscape notably by epigenetic gorges cut across hard rocks which were not apparent at the commencement of dissection. This concept remains perfectly valid for platform regions with an unconformable sedimentary cover. It has little application to geosynclinal chains, apart from regenerated geosynclines. To stress this limitation, therefore, we shall speak of it as *discordant super-imposition*.

In his *Traité de géographie physique*, E. de Martonne demonstrated correctly that the Davisian scheme is poorly applicable to overthrust nappes. When a stream progressively incises itself into such structures it is not capable of adapting to them: they are too variable in the vertical plane as a result of tectonic deformation. But, in this case, the very concept of super-imposition is involved. It implies that the drainage network developed straight away on an initial surface, re-excavating the deep structures. The evolution of the geosyncline is such that this surface is rarely one of erosion cutting across the structure. In almost all cases there is no initial surface but a landscape already undergoing dissection which suffers continual modification arising as much from the dissection itself as from the tectonic movements which continue to deform, uplift, overturn, fold and fracture it. Nappes are exposed at the surface and *décollements* are produced in the cover rocks such that morphogenesis is stimulated. This is the case in the Jura near Lons le Saulnier where Mesozoic rocks have been thrust over the Pliocene marls of La Bresse, with a displacement of between 2 and 3 km.

Therefore, in the Pliocene, the Jura was already dissected by a river system whose evolution was only interrupted, and then not everywhere, by the Quaternary glacial advances.

To designate the poor adaptations which result from this gradual incision of the drainage network into discordant structures (and which, moreover, very often continue to deform during the incision itself) we use the expression *progressive superimposition*.

Progressive superimposition is the normal case in both geosynclinal and superficial fold zones. In effect, the very formation of folds implies discordance. It demands that the strata interact one with another such that there is necessarily an increasing compression of beds with depth on the anticlines and increasing extension in the synclines. Folding in the cover rocks can only develop by a *décollement* in the thrust cover rocks themselves: all the more reason for splinters, overthrusts and thrust nappes.

(*b*) *Antecedence*. The stream, already established, is affected by an interruption of the cycle, by a rejuvenation initiated by new tectonic movements. For example, a fault block is uplifted across the line of the stream. It incises itself and thus maintains its course. Actual examples of such a development are known. For example, to the south of Titicaca the River Huenque, which flows down the slopes of volcanoes forming the watershed, had by the Pliocene cut a valley into the folded Mesozoic cover rocks. It became fossilised in the lower Quaternary by hydrovolcanic tuffs. These tuffs, in their turn, have been folded in a somewhat different style, namely rather massive undulations cut by local fractures. The river has broadly maintained its course, partly exhuming its former valley. But the folding carried on into a period of accumulation arising from climatic change, probably coinciding with the Mindelian. At this time, the anticlines were incompletely worn down and they continued to form obstacles such that lakes persisted in the intervening synclines where deltas were constructed. With the slowing down of the folding and the change in style of the deformation (mass movements and faulting) incision of the anticlines ended in complete removal, allowing the draining of the lakes. The resulting climatic terrace, dating from the penultimate glaciation, is continuous across anticlines and synclines, simply warped where it traverses sharp breaks of slope. There is therefore permanent interference between dissection and tectogenesis, each having its own rate of development variable with time, with the result that equilibrium between them is itself continually changing. Tectogenesis continued in the Mindelian with a degree of disorganisation of the run-off characterised by the appearance of lakes, the dominance of dissection then becoming re-established with modification of the tectogenesis.

The phenomena associated with antecedence are not accidents in landscape evolution as the Davisians thought. On the contrary, they are the rule in every region of active tectonism, of which the geosynclines form a part. But as tectogenesis fluctuates it is necessary to specify every case of antecedence in terms of the types of deformation and the tectodynamic episode.

The distinction between superimposition and antecedence posed by the Davisians, and which has nourished numerous polemics, is a formal notion somewhat lacking in reality. In fact, antecedence and superimposition exist side by side, and not only in geosynclinal belts: both superimposition across discordant structures, i.e. predominantly progressive superimposition, and antecedence arising from deformation and dissection acting simultaneously. It is only gradually, when the final phase of stabilisation is already well advanced, that the role of antecedence diminishes. Even so, cases of antecedence are common in platform regions where they result from warping, as in the case of the Quaternary uplift of the Vosges and the correlative tilting of Lorraine towards the west and northwest.

The drainage network in geosynclinal belts is characterised by the following apparently contradictory traits:

1. A general adaptation to broad-scale movements, that is to the relative mobility of wedges which occurs as much in the cordilleran phase as in the period of progressive stabilisation. The tectonic asymmetry of the Alps, dating from the Neogene, is still found almost intact in the asymmetry of their drainage pattern. The subsidence of the Po basin has resulted in almost no reorganisation of the drainage. Despite the proximity of a base level particularly favourable to incision due to downwarping, the tributaries of the Po have barely captured the headstreams of the Rhône system which dissect the long slope of the range. Only a few basins in the high Durance have suffered capture, a thousand square kilometres in all! Within the Rhône catchment, most of the streams descend the tectonic slope from the crest of the orogen towards the piedmont. It has taken powerful glacial overdeepening in the Dauphinois facies and the excavation of a lake basin to make the longitudinal trace of the Grésivaudan accord with the lithology and not with the tectonic structure. The development of the drainage system of the Ligurian cordillera is just as illustrative. The same adaptation may be seen in the general trace of the River Huenque which flows from the most uplifted zone, on the water-parting between the Pacific and Titicaca, down to the less uplifted basin of the latter.

2. A general lack of adaptation to local structures (discordance), alternates with adaptations in detail. The antecedent characteristics of the River Huenque within anticlinal structures, for example, are placed in this category of failure to adapt to local structures. But these do not exclude certain detailed adaptations. An anticline may be cut by an angular trace which is guided by faults associated with the flexures. In a general way faults, in shattering the rocks, favour the incision of streams especially when the faults are too recent for the associated breccias to have been recemented. Yet other small-scale adaptations are made in relation to the lithology, but they are more rare and much less persistent than the deformation itself. It is

above all in the calcareous rocks that ill-adapted courses tend to be abandoned, due to the ease of development of subterranean flow independent of subaerial flow. In other rock types, discordant courses are readily maintained, thus minimising the relative importance of a lower base level and so explaining the minor importance of stream capture along the Alpine or Ligurian watershed.

As long as stabilisation of the geosyncline has not advanced too far, general adaptation to block movements and a more local discordance usually exists alongside, for example, antecedence in respect of minor structures, or progressive superimposition. This discordance on a minor scale does not rule out, in its turn, detailed adaptation of drainage to structure, principally in shatter zones. As compartmental movements are not attributable to geosynclines, where they are merely more violent and changeable, the same mechanisms are found again in the other types of structures, in both the folded cover rocks and in the platforms.

Folded cover rocks and their relief

Cover rock folds (*plis de couverture*) are those which affect the sedimentary series of platforms not involved in geosynclinal evolution. The Franco-Swiss Jura is a classic example as are certain pre-Alpine massifs such as the Chartreuse, Vercors and Genevois blocks. They owe their architectural distinctiveness to the nature of their material and to their mode of tectogenesis. In turn, this is strongly expressed in the relief. The relief of the Jura in particular has been taken, since before Davis's time by de La Noë and de Margerie, as an example of the influence of structure on landform.

Folded cover rocks are found in regions which have been subjected to platform sedimentation and which have then suffered tangential pressures. The folded material is made up of beds which are thinner and, above all, more variable than those found in geosynclines. In the Jura, as in Lorraine, alternation of limestones, marls and calcareous marls are found. In the Saharan Atlas about Colomb-Béchar, it is a series of grits and clays. A succession of alternately rigid and plastic beds is the rule, allowing a more or less complete detachment of the *décollement* from the rigid substratum and a differential movement of the beds themselves. The fold style is strongly influenced by the particularly variable mechanical properties of the strata. Accordingly, lithology does not merely control the dissection: it controls the tectonics.

The folds affecting the cover rocks have their origin in tangential movements. They are found in three types of tectonic condition:

1. On certain detached portions of the platforms adjacent to geosynclines which become incorporated in the geosyncline during its evolution, as we have shown above.
2. At the edge of geosynclines whose particularly violent tangential stresses

FIG. 2.16. The flank of an anticline, at southern foot
of Lubéron, western pre-Alps
    Massive Urgonian (lower Cretaceous) limestones,
bent into an upfold and crossed by a small gorge.
Several fissures appear in the massive limestone,
which is overlain by thinner beds, less resistant and
partly eroded. The anticlinal flank has become a
structural surface

are transmitted to the adjacent platform. This is the case in the Pre-Alps or
the Jura and in the Saharan Atlas of Algeria. Of course, this occurs particularly in the intracontinental geosynclines. Examples abound in the
Mesogeic system but are rarer in the circumpacific system.
3. Even in the interior of platform regions, great fractures have made possible tangential block movements and compression of a sort. This is the case
in central Germany, in small chains of the Weser district, in Palestine, in
Libya and in Syria, and in flexures running along the extremities of the
great trench of the Red Sea.

    These diverse types of occurrence do not prevent folded cover rocks from
having common architectural characteristics which arise directly from the

nature of the material, in particular from the degree to which the cover becomes detached by *décollement* from the rigid basement below. Accordingly, we shall consider the various tectonic styles and the corresponding forms in conjunction.

Flowing parallel folds (classic Jura style)

Following the works of de La Noë and de Margerie, the Jura has been taken since the end of the nineteenth century as the type area for the control of relief folds of flowing style (*plis souples*) in cover rocks. From that time the expression 'Jurassian relief' has been used to define the forms derived from parallel, symmetrical superficial folds. In fact, this concept corresponds only partly to reality. More recent detailed studies have shown that such folds are rare in the Jura and limited to certain parts of the range in Switzerland and close to the Franco-Swiss frontier. The fact is that before the phase of folding, the Jura was subjected to faulting in the Oligocene, an extension of the faulting of Alsace, the plains of the Saône and the Bourguignonne scarplands. The folds became superimposed on the faults as has been clearly demonstrated in France by L. Glangeaud and his pupils and M. Dreyfuss and the Besançon geological school. Following them, we shall designate the tectonic style characterised by superimposition of folds on faults the Comtois style (after Franche Comté). It will be discussed below.

In the eastern Jura, the basement is covered by a thicker sedimentary cover. In fact, up to the orogenic phase of the Alpine geosyncline, i.e. as late

FIG. 2.17. Anticline in quartzitic sandstones, in the Venezuelan Andes
A surface-fold in rigid rocks. The anticlinal folding is accompanied by crushing in the heart of the fold and a fault on its left flank.

as the Oligocene, the region was inclined towards the geosyncline in the east and southeast. The sediments thicken in this direction, and moreover, they have suffered little erosion in subaerial periods while considerable denudation was effected in the western Jura. In the stronger sedimentary assemblage of the eastern Jura faulting was less important than in the cover rocks cut up by erosion of the comtois Jura. Folding has occurred just as readily, but opposition of folds and fractures is less obvious. It exists, nevertheless, with the result that the Jura is not a very typical example, though regarded as the classic case. The Atlas of the Algerian Sahara is more typical of the model. For all that, the eastern Jura can be considered the type: faults bound basement blocks on which typical superficial folding has developed.

TECTONIC STYLE OF FLOWING FOLDS

The sedimentary strata covering the platforms favour development of the flowing type of fold for two reasons: they are well stratified and made up of successive layers which are usually rather thin; and they are made up of alternating rigid and plastic beds, such as grits or limestones and marls or clays respectively. It follows that when affected by folding such bedded materials respond differentially to one another and undergo relative displacement. The thinly bedded calcareous marls with marl intercalations are particularly favourable to this kind of deformation, exactly as a ream of paper can be folded by hand more easily than an equal thickness of cardboard. The plastic beds are thin in zones of tension and thicken under compression. Such adjustments delay any tendency to fracture in the beds and allow the persistence of a flowing symmetrical fold style even under severe tectonic stress. In contrast, the rigid strata, especially quartzites and grits occurring mainly in thick beds, fracture readily: in this case, the symmetrical fold style is rarer. Nevertheless, it is represented in the Saharan Atlas of Algeria in a series where sandstones alternate with clays or marls; these plastic members compensate for the rigidity of the grits in a region where folding has not been very violent.

Folds of the symmetrical style are not generally found in isolation. Several anticlines and synclines are usually aligned side by side running all in one direction. They make up a sheaf of folds. In such cases, the folds are always elongated, clearly longer than they are broad. But sometimes, mainly in areas with conflicting tectonic trends, the lengths of folds may not exceed twice their breadth. Then we speak of *domes* (anticlinal) and *basins* (synclinal). While the folds of the northern Pre-Alps and the Jura, controlled by the Alpine orogeny alone, are arranged in a group, those of the southern Alps from the town of Gap southwards (where this orogeny conflicts with the older Pyrenean–Provençal folding) frequently contain domes, basins and arcuate folds.

In a sheaf of folds, each fold terminates after a certain distance. Frequently, a syncline replaces an anticline on the same axis and vice versa.

FIG. 2.18. Anticlinal ridge crossed by a gorge, Kuh-i-Kialan, Iran (drawn by Rimbert from an aerial photograph)
   A pericline in limestone creates a ridge by the erosion of the marls and evaporites that form the valley beyond it and make a ring round it. Streams, superimposed from the evaporite-marls, have cut gorges, notably in the foreground and further back, just before the pitch of the pericline takes the limestone beneath the marls

Then we speak of reverse folding. The crest of the anticline drops progressively to pass into the trench of the syncline. We speak of periclinal termination or plunging pericline if considered from the point of view of the anticlines, or of remounted perisyncline if considered from the point of view of the synclines. In both cases, the dips change direction and become divergent. The anticlines do not always form perfectly regular whalebacks, for often their crest drops locally to rise again: this is termed a saddle.

   The style displayed by flowing parallel folds varies as a function of the manner of tangential pressure on the one hand, and the disposition of the beds on the other. When the pressure is violent and directed towards a zone which is depressed, be it due to downwarping or overturning, the folds become asymmetrical. The flank facing the depressed region is all the steeper the more marked the depression. Then asymmetrical folds pass into recumbent folds which usually accompany box-folded anticlines (see

FIG. 2.19. Mountainside, and part of annular valley, near Bienne in the Swiss Jura
    Cretaceous rocks giving purely structural forms without any detectable planation surfaces. The mountainside, steep and wooded, is a structural surface of limestone. In the centre is a marl outcrop, eroded into a cultivated valley that forms a ring round the limestone hill on the left

below, p. 142). The Vercors arch in the Valence Basin is an excellent example of a regional recumbent fold extending for over 30 km. Asymmetrical folds may even lie one on top of the other. We speak of recumbent folds when, along the same vertical axis, both flanks of the anticline are met with in succession.

According to the manner in which they were pushed and the disposition of the beds, the relative size of anticlines and synclines varies. Moreover, when the relative dimensions become too unequal, we pass from the flowing fold style to a fold style typical of the more rigid cover rocks which we study later (p. 141). Deep and narrow synclines are said to be *pinched synclines*. In contrast, broad synclines lying between narrow, crushed and often asymmetrical anticlines, give an ejective style to which we apply the term *contained synclines (synclinaux en baquet)*.

Of course, such diverse arrangements of folding have repercussions on relief forms. That is why they cannot be ignored in geomorphology.

RELIEF ON FLOWING FOLDS

Jura-type relief is characterised by flowing superficial folds from which relief forms are derived both directly (relief conforms to the structure), and indirectly by means of inversion of relief.

Jura-type relief forms

These owe their geomorphological definition to de La Noë and de Margerie. An anticline with a crest of resistant rock preserved is called a *mount*. When this crest has been stripped of any superjacent weaker beds we speak of a *derived mount*. In a concordant stratigraphic assemblage the derived mount is made up of a structural surface. It is a tectostatic, differential erosional form. On the other hand, when the anticlinal crest is made up of superficial beds which crop out at the time of folding, we speak of a *primitive* or *original mount*. It constitutes a tectonic form, be it still in course of folding and therefore an active tectonic form, or recently created by folding and not yet demolished and therefore a residual tectonic form.

Of course, the mountain forms match the type of deformation, and the asymmetry of the folds is reflected in the asymmetry of the relief, at least up to a point. In fact, as we shall see, it is rare for the flank of an anticline to be preserved when the dip is very great.

On the flank of a derived mount, one often sees the remains of the unresistant bed whose removal has allowed the stripping of the structural surface now constituting the hill. There frequently rests on this weak stratum a resistant bed preserved especially in the neighbouring syncline. This alternation of hard–soft–hard strata gives rise to homoclines on the

Fig. 2.20. The valley of Pont de Gueydan, Alpes Maritimes

A sharp syncline, with steep flanks, from which softer rocks have been largely removed (though still present in the foreground, under the railway). The up-tilted strata, on the right, form a structural surface, through which the river Var has cut a gorge. The periclinal end of the syncline, in the distance, takes on the form of a boat

flank of the mount, which the descending torrents dissect into *herring-bones* or the *flat-irons* of the American writers. The repeated alternations of marls or clays with limestones or sandstones particularly favours the emergence of flat-irons in climates conducive to intense stream action.

The action of the stream network on the mountains produces certain recognised forms. Some rivers cut across the mountains. Because of the nature of the relief, valleys become particularly deep and narrow where they cross resistant strata. The stream may form a gorge, spoken of as a transverse or discordant valley. Transverse valleys frequently adapt themselves to the local structures, cols or minor fractures. But this does not solve the problem of their origin, which can be discussed validly only at the regional scale by reconstructing the evolution of the mountain range or, at the very least, a large part of it. The gashes cut by local torrents into the flanks of the anticline but which fail to reach its crest, form dipslope valleys (*ruz*). However, the majority of such valleys owe their origin to particular morpho-climatic conditions and not to 'normal erosion' by temperate latitude streams. In the Jura, they are rare and are generally formed under periglacial conditions, for example under the influence of nivation. The semi-arid lands are more favourable to their development, due to severe stream action in violent stormy rains which do not have time to infiltrate the regolith and so flow over the surface. Such dipslope valleys are numerous on the flank of the Negev anticlines in Israel.

Ridges are complementary to vales. A vale, in effect, is a depression coinciding with a syncline. Except in very rare cases, the beds preserved in the vale are more recent than those which crop out on the ridges, implying a differential denudation guided by the fold structure. In the lower country making up the vale, denudation is less intense than on the neighbouring ridges. When the lithology is favourable, this allows the development of flat-irons on the flanks of the ridges, indeed of homoclines forming the peripheral crest of the synclines with their cornices facing outwards up the dip. These are *external crests*, similar to those which surround perched synclines, but without having suffered inversion of relief. The syncline, being lowered topographically more rapidly than the neighbouring anticline, remains a vale.

Inversions of relief are characterised by the fact that the topographical forms are the opposite of those of the structure. These inversions are found in various degrees of development. An attenuated variety of relief inversion is provided by the anticlinal strike valleys (*combes*). These are depressions which breach the crest of the anticline with the result that the most elevated part of the anticlinal ridge is not on its structural axis but on its flanks. In the development of relief, the strike valley coincides with the appearance of external crests on the margins of a vale. It is dominated by the homoclines formed by the resistant stratum of the partly eroded anticline, beneath which the less resistant beds (generally of marls or clays, exceptionally of little-consolidated sandstones beneath quartzitic sandstones) have been reached. The updip brow of the homocline faces the anticlinal axis un-roofed by erosion.

An extreme kind of inversion of relief is offered by regions with synclinal mountains and anticlinal valleys. The troughs of the synclines make up the mountain summits with the most elevated limbs of the synclines marking the perimeter and coinciding with the summits of the external crests. At their foot, the anticlines are depressions, generally made up of ridges of clay or marl. This is the situation in the Forest of Saou and on several mountains in the Laragne region, in the southern Pre-Alps. (See geological map *Privas*, 1 : 80 000 scale, for the Forest of Saou; and *Le Buis* for the environs of Laragne.)

FIG. 2.21. An asymmetric anticline with eroded core, Bou Amrane, in the middle Moulouya valley of Morocco
On the left, the gentle limb gives a marked homoclinal ridge facing the valley. On the right, the steeper limb has been more eroded. There is a girdle of scree-slopes, more or less dissected

Genesis of Jura-type relief

These diverse relief forms have been grouped together in an evolutionary series by the followers of Davis. A relief of folded structure appears as a result of postulated severe and rapid folding and its immediate geomorphological expression. This is an original relief of ridges and vales. Erosion attacks the hills more vigorously, resulting in dipslope valleys on their flanks. At their summits, the dipslope valleys join together and strike valleys appear. As the latter develop, the anticlines end up by being eroded to such a degree that the synclines stand above them. The ultimate condition is then reached: inversion of relief is pushed to the extreme with the appearance of perched synclines.

Unfortunately, such an evolution has never been demonstrated and is purely a product of the imagination, to which W. M. Davis made too eloquent an appeal. Every time *plis de couverture* have been studied, evidence of the interplay of morphogenesis and tectogenesis has emerged. When these folds develop in a limited paroxysm, when tangential movements are particularly intense, they do not emerge ready made. Such paroxysms always last some millions of years, commonly between five and ten, as in the case of the folding subsequent to the molasse of the Pre-Alps and of the eastern Jura. Now, with Davis it is a question of short paroxysms. It is quite

FIG. 2.22. Eroded anticline with core-ridge, Suchet, in the Jura
On the right, the massive mountain with the woods and pastures represents the gentler flank of the upfold; the beds dip to the right. The anticline is asymmetrical, with a steeper left flank and an eroded core that is seen left centre of the photograph. Within the valley the rounded hills form a *mont dérivé*, due to circum-denudation. The limestone scarp on the left, in steeply-dipping rocks, forms the sharp edge of the Suchet. The corresponding scarp on the gently-dipping right side, is much less prominent.

evident that morphogenesis is not interrupted during earth movements and that it acts on the strata even while they are being folded. The interplay between tectogenesis and morphogenesis begins as early as the commencement of folding. The persistence of morphological forms renders this an inevitable conclusion.

For example, in the eastern Jura the anticlines are usually of Cretaceous material, while the molasse persists in the synclines. But the molasse does

FIG. 2.23. Outer crest of a 'perched' syncline, on the western edge of the
Forest of Saou (French 1:80 000 series sheet 199, Privas)
  A radial fault breaks the crest-line of the scarp on the left

FIG. 2.24. Discordance on the flank of an anticline (southern flank of Lance
anticline in the pre-Alps of Diois)
  On the left, the limb of the anticline, formed by the chalky limestones of
the upper Jurassic, giving inward-facing scarps. The beds are truncated by a
warped erosion surface. On the right, a scarp of molasse, unconformable on the
Cretaceous and less folded. The very flat surface of the molasse plateau is an
old Quaternary slope that truncates the molasse. In the foreground is
another more recent *glacis*.

not lie conformably on the Mesozoic. A long period of submergence is evident, from the end of the Cretaceous into the Oligocene, during which both deformation and denudation occurred. It appears that the strata were in a mildly folded condition before the Oligocene. In every case, the molasse re-covered beds within the synclines younger than those forming the anticlines, as if the undulations had been cut across by a pre-molassic surface. This cover, thicker in the synclines, successfully survived the tangential movements of the Miocene. Conversely, being much thinner over the anticlines, it has facilitated their uplift. Generally, therefore, folding has been perpetuated *in situ*. But the flanks of the hills are not structural surfaces. Around Pontarlier, for example, it is clearly a case of an erosion surface having been newly folded, for it truncates older and older beds towards the ridge summits and, conversely, progressively younger strata towards the synclinal axes: here the Cretaceous has been preserved with the middle Jurassic making up the hilltops. With such hills and valleys, we may speak of tectonic relief with the discordant molasse more or less cleaned off, or of original relief: this is pre-molassic and is expressed in the morphogenesis only through the intermediary of tectonics.

On the anticlines, the infra-molassic erosion surface often cuts across marly members such as the Portlandian or the Oxfordian. In such a case, differential erosion proceeds as follows. A strike valley develops on the

FIG. 2.25. Production of ridges by renewed uplift of an anticline, in the region of Chota, Peru

The end of an anticline, showing clearly the periclinal disposition of the strata. The alternations of marl and limestone have plunged under the marls and clays in the foreground. The anticline has been truncated by planation and then subjected to renewed uplift. Differential dissection has produced homoclinal ridges along the edges of the harder beds

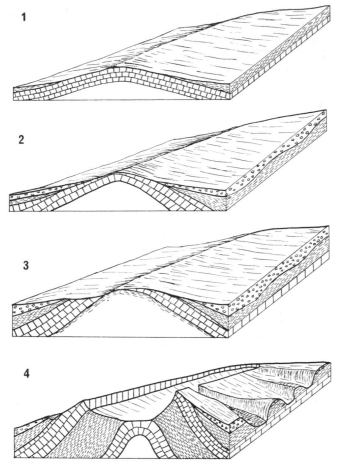

FIG. 2.26. Formation of an eroded anticline

1. The upfold starts in a series of alternating limestones and marls. As the folding increases, the overlying marls are eroded and a low limestone ridge is formed.

2. The continuation of folding entails the decapitation of the anticlinal arch, and mechanical attack on the limestone to produce a debris-covered glacis on the outer slopes. (It is assumed that the climate is semi-arid, permitting the mechanical weathering of the limestone and the spreading of debris on the slopes.)

3. Further uplift and erosion have removed the limestone from the fold-axis, exposing the marls underneath.

4. A renewal of folding accentuates the arch, particularly since the crest of rigid limestone has already partly disappeared. The conglomerates on the glacis are themselves deformed and dissected. On the marl outcrop, a combe-like valley becomes wider with the retreat of the inward-facing cliff-scarps. Another limestone bed, below the marls, begins to be exposed; ultimately it may form a *mont dérivé* (cf. Fig. 2.22)

crest of the anticline where the weak strata crop out. This is the normal kind of initiation of strike valleys on symmetrical folds, but in humid climates, there is no reason at all why the anticlinal crests should be attacked more rapidly than the bottoms of the synclines. After all, the beds are in a condition of extension, and their joints and fissures are consequently enlarged: thus, water infiltrates readily and reappears lower down, on the edge of the syncline. On limestone rocks, this process produces a subsurface solution and karstic forms without, however, giving rise to strike valleys. It is for this reason that dipslope valleys are exceptional, and torrential gullies negligible on the flanks of the Jurassic mountains; rather than flowing in streams, the water infiltrates. It takes a semi-arid climate to modify these conditions, for in such regions, dipslope valleys are much more common.

Perched synclines are explained in quite another manner. In a region such as the Larange, a substantial bed of black marls of middle Jurassic age is surmounted by limestones. Domes and basins were formed in the first earth movements. Then, during a long period of denudation, the beds were truncated. The limestone was strongly dissected on the anticlines but persisted within the synclines. The heterogeneity of the resulting folding has influenced the course of the renewed tangential movements. On the anticlines, where there is little rigid limestone but abundant marl, the beds have been readily folded. Here, complicated folds have developed and have dislocated the dissected limestone. Such areas have been easily eroded. In the

Fig. 2.27. Relief developed on folds, in the Chota region of Peru
Alternations of folded Mesozoic limestones and marls, cut by an erosion surface and further uplifted by a gentle anticlinal warping that has provoked differential erosion and the incision of streams. The limestone beds stand out as homoclinal ridges, and there is a conforming land-use pattern

synclines, on the contrary, the limestones have been preserved by their great thickness: they have resisted erosion and more recent earth movements. They have been compressed into rather tighter folds; this has bolstered their resistance and allowed their gradual isolation into perched synclines.

Such polygenetic developments seem to be the rule in all styles of folded sedimentary rocks. An initial earth movement creates tectonic relief which is simultaneously attacked by erosion. As the beds are progressively deformed, they are modified by denudation. Debris accumulates in gorges, rendering them resistant to folding. The planation or incision of anticlines, on the other hand, reduces their resistance to the same folding and so facilitates destruction of hinge lines (*charnières*). The process is particularly clear in the ejective fold style. The *contained synclines* are usually made up of accumulations of rigid, little folded beds very resistant to tangential movements. In contrast, the anticlines which punctuate the synclinal masses are responsible for the uplift of the more plastic subjacent beds. When compression intensifies, the rigid beds of the synclines form a cap and a veritable ejection of the subjacent plastic material is produced within the anticline, thus giving the name to this style of folding. The resulting inversion of relief is quite evident. The synclinal massifs, in which the rocks have retained their compactness, are resistant to erosion. The anticlines, on the other hand, being composed of ejected plastic material or dislocated remnants of more rigid beds, are readily excavated.

Finally, the formation of transverse valleys (*cluses*) may be understood. The explanation involves a combination of mechanisms already analysed in relation to the geosynclinal chains. In the transverse valleys of the high Jura one finds a conjunction of (*a*) a reach which is antecedent with respect to the deformed inframolassic surface on which streams flowed and which was inclined towards the molassic sea before the surface became folded in the most recent earth movements; (*b*) a superimposed section within discordant structures beneath the molasse, notably within those folds displaying a certain disharmony even when they are regular; and (*c*) a section showing adaptation to diverse structures, notably to faults which antedate the recent folding and which have been dislocated and offset. To the south of Pontarlier, for example, the Ronde Fontaine stream and then the River Doubs utilise such a structure.

This combination of tectogenesis and morphogenesis is found in the interaction of folds and dislocations. It is because of this interaction that the term morphotectonic has been coined by geologists.

Interaction of folds and faults

Two fundamental cases will be considered. In certain regions, for example the Comtois Jura, some major faults cut up the region during the first tectonic phase. The folds were then produced: they have interfered with

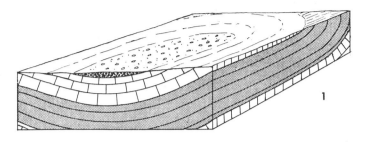

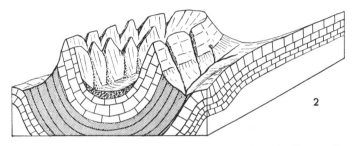

FIG. 2.28. Origin of a 'perched' syncline (based on the Forest of
Saou—see French 1:80 000 sheet 199, Privas)

1. An early folding phase has given a synclinal trough of
middle and upper Cretaceous strata, in the centre of which are
sandy detrital deposits of Eocene age.

2. At the present time, after renewed folding in the Miocene.
The hard outcrops of upper Cretaceous limestone have been bent
into a narrow bucket-shaped hollow with near-vertical sides.
They form strong crestlines, cut into chevrons by ravines and
boulder-screes. The middle Cretaceous marls have been largely
eroded on the flanks of the neighbouring anticlines, creating an
inversion of relief. Underlying limestones, on the right, form an
anticlinal fold that is becoming a *mont dérivé*

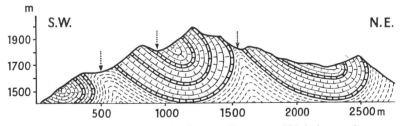

FIG. 2.29. Perched synclines in the Basque Pyrenees of Binbalèta and
Kartchila (*after Viers, 1960, p. 101*)

Paleocene limestones rest on Maastrichtian slates. The folded complex,
with narrow anticlines pinched up between broad synclines, has been
further uplifted. The limestones of the synclines, being more resistant, form
the summits, but the effect is lessened by the narrowness of the intervening
upfolds

the faults and this has modified the style of folding. In other regions, the polygenetic evolution is characterised, throughout its development, by fold dynamics, as in the Pre-Alps of Grasse. But the interplay between tectogenesis and morphogenesis results in the appearance of faults within the folds which are then transformed into fault splinters in a later earth movement.

THE COMTOIS STYLE

This style is represented in the Jura to the west of the Franco-Swiss frontier by progressively thinning cover rocks which have been cut across more deeply by the mid-Tertiary erosion surface, thus rendering the cover thinner and more rigid. Here the faults have played a decisive role. They are later than the folds and cut across them in a north–south and southwest–northeast direction, like the structures in the neighbouring Hercynian platform (plains of the Saône, Alsace and Bourguignonne scarplands). They date from the Oligocene and so antedate the upper Tertiary folds.

The distinctiveness of the comtois style rests in this fact: the faults have created mechanical discontinuities in the cover rocks, discontinuities which, in turn, have influenced the folds. The comtois style is defined therefore as a style characterised by the superimposition of folds on faults. Some types of original deformation arising from this have been analysed by L. Glangeaud and his school. In the most simple case, such as occurs to the south of Pontarlier, the folds correspond to a tangential compression orientated parallel to the direction of the faults (see geological map *Pontarlier*, scale 1:50 000). The *décollement* of the cover rocks and the folds which have resulted vary along the faults which are responsible for the mechanical discontinuity. In the eastern section, the beds have moved further. Therefore, the fault has been transformed during dislocation due to a progressive change from a dominantly vertical stress to one dominantly tangential. The consequences on the relief are as follows:

1. The fault plane, a shatter belt, constitutes a line of weakness which has been exploited by erosion. Here a valley has been cut from which strike valleys have been etched into the crest of the anticlines where this valley cuts across the weak beds of the fold axis. But the strike valleys scarcely form spring points and sapping is limited.

2. The folds are offset from one part of the fracture to another. The anticlinal and synclinal axes do not correspond, and neither do the relief elements. As a result, they are disposed *en echelon*. Further, the drag along the fault plane has often resulted in an upturning of the margins of the anticlines which become recurved. All this is reflected directly in the form of mountains, valleys and strike depressions.

In the plateau Jura, things are more complex because the faults are oblique or perpendicular to the direction of the tangential stress. The following types of discontinuity may be distinguished:

The fold-fault

A fault trace very nearly perpendicular to the tangential stress is dislocated by the folding strain. It is folded in the same direction as the thrust. If the most elevated topographical element lies towards the direction from which the push came, this mass overrides the other. In effect, the beds making up the fault escarpment have below them a depression: there is no obstacle in front of them and the thrust works freely. In contrast, below the level of the floor of the depression the beds come up against the strata of the adjacent structural unit and this limits their movement. As they are sufficiently plastic, the strata of the uppermost block recurve in a half anticline on the trace of the contorted fault. As they are now being extended, the strata begin to fragment and become fragile. They are now more readily denuded and, after a little while, on the most elevated mass where the two sets of strata are superposed, the fault gives rise to ridges with, at times, linear depressions along the weak strata and crests along the harder beds (Fig. 2.30).

Pinch

During the tectonic faulting phase, tension develops and a wedge-shaped mass subsides between two faults which converge at depth. Folding intervenes and is expressed as a compression. This wedge, more fragile than the two more massive compartments which encase it, becomes crushed. Its component beds are violently folded. If they are plastic, they are drawn out in an anticline whose axis inclines in the direction of the push. In detail, the beds are very contorted and broken up, which diminishes the resistance of coherent facies. Such anticlines never give rise to mountains, at the very most only to heights without any structural arch. They tend to produce inversion of relief. If such inversion is not always realised in the Jura, this is only because insufficient time has elapsed. It will be remembered that due to *décollement*, the Mesozoic cover rocks of the Jura overlap the Pliocene of the Bresse. Still, so long as compression continues, and it appears that this phenomenon has not yet terminated, the anticline of the pinch continues to grow, so compensating downwasting and, occasionally, more than just compensating for it.

Anchorage

This also results from heterogeneity of beds from one part of a fault zone to another. The principal faults affect not only the cover rocks but the shield beneath. When the face of a master fault is disposed tangentially to the direction of tectonic push it blocks the *décollement* of the cover rocks like a wall stopping a sliding carpet; and if the push continues, makes it fold. Thus an anchorage is produced. It is as though the cover rocks are anchored to the obstacle which prevents their progression. The cover becomes folded upon itself against the obstacle, and an arch is produced on the margins of

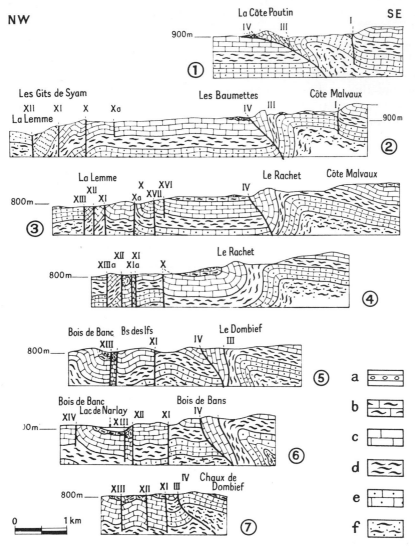

FIG. 2.30. Jurassic folds. A series of cross-sections in the Syam area, east of the plateau of Champagnole, in the 'plateau Jura' *(from Guillaume, 1961, p. 108)*

a. Tertiary; b. Cretaceous; c. Portlandian–Rauracian; d. Argovian–Oxfordian; e. Middle Jurassic; f. Lias and upper Trias.

A faulted area situated on the edge of two structural compartments has been subjected to compression, causing reversed faulting (as in Section 7), passing into recumbent folds with over-thrusting (as in Section 1). The evolution of a monoclinal fold into a fold-fault, with or without over-thrusting, is well seen on the eastern side of the sections, starting with Section 4 and moving to Section 1 and Section 7 the downthrown block. The most important chains of the tabular Jura (Jura du Plateau) are due to the development of anchorages on the master faults. However, they can also be produced, not only at the level of the shield, but within the cover rocks themselves. All that is needed is that beds be disposed along the length of the fault in such a manner that a plastic mass on the side which is pushed is opposed by a rigid mass at the same level. This plastic mass crams itself against the fault and gives rise to an anticlinal swelling (boursoufflement) (Fig. 2.32).

Anticlines produced by this process are always complex and violently dislocated. Generally they have only one flank, the other being replaced by the more or less recurved fault plane. Also they are readily broken up by erosion, often at the same time as they are being formed. Generally, the clearest forms are monoclines in the resistant beds of the flank, where they are a sufficient distance from the fault to have escaped any considerable crushing.

FIG. 2.31. Eroded monocline at Plaisians (Drôme), in the Baronnies area of the pre-Alps. On the right, a high and massive homoclinal ridge, made of resistant Tithonian (upper Jurassic) limestones. These are displaced toward the left in a fault splinter; this has fissured them and prevented the development of a prominent scarp crest. It is the gentler limb of the monocline overlapping the steeper. In the foreground, the massive Urgonian (lower Cretaceous) limestones produce a sharp *barre*, cut by V-shaped gorges. The vertical strata belong to the steeper overthrust flank of the fold. In the immediate foreground there are vertically-dipping calcareous marls in which a broad, cultivated valley has been eroded. Between the Urgonian limestone ridge and the Tithonian limestone homocline, the core of the fold has been excavated to form an anticlinal valley of great asymmetry. The Jurassic marls and clayey limestones produce gentle, well-cultivated slopes which pass into the depression behind the *barre*

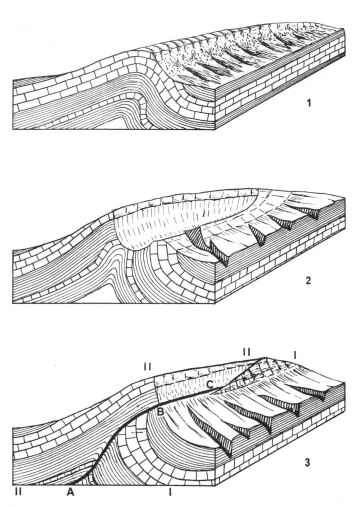

Fig. 2.32. Formation of an overthrust. Based on examples from Provence (on the
1:80 000 sheet 225, Nice)

1. In the first phase, an asymmetric anticline (or monocline in English usage)
is formed. The limestones of the arch are cracked by the tension, and are thus
more easily eroded on the steep limb, which forms a cliff-scarp with scree-cones
at its base.

2. The folding intensifies, and is aided by the previous thinning of the
arch-crest. At the same time, the axis is greatly eroded, and a valley occupies it.

3. Vigorous tangential forces push from left to right. Where the limestone
crest has been replaced by a *combe*, the resistance is least. The gentle flank of the
anticline (on the left), which is more rigid, is better able to resist the pressure, and
it moves forward over the *combe*, whilst the right flank is bent under. At the
further end of the overthrust, there was no valley, for the fold ended periclinally.
The anticline breaks and turns over on itself without overthrusting. Here there is
a mass of plicated limestone sticking up above the marls. In the foreground, on
the other hand, the overthrust beds form a fine homoclinal ridge

The comtois style is thus characterised by an extremely complex tectonism : this is one which changes over very short distances and is very different from traditional theoretical views on 'Jurassic' relief.

THE PROVENÇAL STYLE

This tectonic style is characterised by a succession of several major earth movements all of which produce folds and which interfere with morphogenesis. Accordingly, these are folded relief forms which are folded anew in the course of successive earth movements. Provence provides numerous illustrative examples, admirably studied by that great pioneer worker Lutaud. But the style is very widespread and occurs even in the Swiss Jura, notably in the Grenchenberg (Bernois Jura).

At first, an anticline begins by forming a modest undulation, which is truncated by an erosion surface. In the course of a new tangential push, the fold re-forms on the same spot, the attenuated arch being more readily deformed. Accordingly, the anticline becomes accentuated. It is attacked by erosion and hollowed out to make a valley. However, if the tangential push is accentuated anew, this valley modifies the mechanical behaviour of the fold. In the face of the hard and rigid bed which forms the crest on the up pressure side, there is no resistance, only the *combe*. At depth, in contrast, the continuity of the beds provides a degree of resistance. Structure apart,

FIG. 2.33. Barrier ridge (*barre*) at the foot of the Vercors, in the pre-Alps (La Baume Cornillane)
Vertically tilted beds of Oligocene limestones and conglomerates at the base of the monocline bordering the Vercors. The bed is cut across by some *glacis* punctuated by minor residual relief forms. It is flanked by marls in process of removal (steep-sided gorge in centre of picture) to produce the *barre*

we are reminded here of a similar case, that of the fold fault. The summit on the pressure side overlaps the *combe* and even the summit in front of it. In fact a shear is readily produced in the weak beds which allowed the excavation of the *combe*. The symmetrical anticline gives place to a thrust mass which then evolves into a homocline.

This kind of evolution is favoured by asymmetrical folds, particularly those of the recumbent type. In fact, on a strongly asymmetrical fold whose two flanks dip for example at 45° and 10°, a steeper flank is particularly susceptible to erosion. Under tension, the beds are distended and more fragmented. Incidentally, the steep slope stimulates excavation which further augments this fragmentation. Screes become well developed, while the streams arising from the severe slopes are little inhibited by infiltration. In contrast, the gentle flank resists erosion very well like a monoclinal back-slope. The steep flank is also rapidly cut up with the growth of a flanking strike valley (*combe de flanc*), which forms an asymmetrical depression below the gently dipping limb which has become detached to form a homocline. Such a disposition considerably facilitates shearing and overlapping, given a renewed accentuation of pressure from the same direction. The homocline of the gentle flank is then transformed into a thrust mass which overlaps the *combe de flanc* and the upstanding beds of the steep flank. To the north of Grasse, the Pre-Alps in the district of Nice provide numerous particularly illustrative examples. Here, where the beds are less steeply dipping and have not been eroded, the overlaps terminate at the extremities of the anticlines. The overlap ends precisely on the edge of the *combe de flanc*, being at its greatest where the *combe* was broadest (see geological map, *Nice*, scale 1:80 000). More severe erosion of the steep flank of an asymmetrical anticline can produce an accentuation of its asymmetry during later earth movements, without going so far as to give rise to a thrust block. This is true of the Luberon area to be seen on the 1:80 000 Forcalquier map. In contrast, a very accentuated recumbent fold can give rise to an overthrust fold following a major *décollement*, allowing the beds of the upper part of the fold to move across those of the lower. Such an evolution appears to have begun in the Pre-Alps of the Niçoise between Villefranche and Menton.

Successive episodes of folding can introduce the most varied complications into the folds of the sedimentary cover. This is the case with tectonic inversions. In a tectonic inversion, the same unit that evolved as a syncline in an initial paroxysm can develop into an anticline in a second period of earth movement. This has been the case in the Djebel Batène in Tunisia. It is obvious that this type of tectogenesis is favourable to the formation of perched synclines. This is a long way from the classic explanation of the development of 'Jurassic' folds, but the error is readily explained by (*a*) an erroneous starting point, namely assuming readjustment in the time between tectogenesis and morphogenesis, leading to a catastrophist conception of tectodynamics; (*b*) a one-sided, narrowly specialist point of view, thinking only in terms of morphogenesis instead of placing it

FIG. 2.34. A homoclinal scarp formed by the front edge of a thrust mass, at Lac de la Motte in the Jura
Portlandian (upper Jurassic) limestones are thrust over Valanginian (Lower Cretaceous) chalk-marls. They dip at about 10° E, giving a cliff-face, that is somewhat irregular because of the fracturing of the limestone. There are blocks irregularly inclined, and stacks of boulders along the crush-zones

in the vaster framework of the whole dynamic ensemble of surface pro-cesses; (*c*) a faulty method: to imagine explanations instead of patiently reconstructing the evolution of a region with recourse to all available methods. A lesson is implied here.

Folds in rigid cover rocks

We have seen cases in which the cover is always detached from the rigid shield at depth thanks to the presence above the shield of plastic beds, generally marls and clays which are saline to some degree, like the Trias of the Jura and of Provence. However, some series of cover rocks are made up of rigid beds which are present in some considerable thicknesses on the surface of the shield, e.g. limestones and sandstones. In such cases, no *décollement* of any importance is possible, and only phenomena of rather limited discordance appear. These are the rigid cover folds.

The minor importance of intercalations of supple beds in this type of structure limits the action of lithological influence in morphogenesis. Inversions of relief are exceptional, conformable relief being the rule. Rare mobile beds, usually of little thickness, give rise everywhere to chevron folds and little monoclines on the flanks of folds. Neither erosional chasms nor perched synclines are produced. The relief is essentially tectonic.

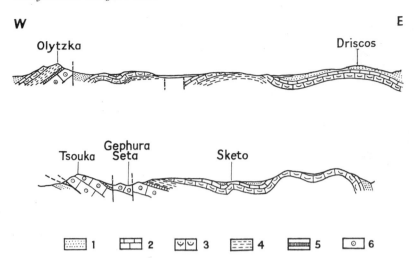

FIG. 2.35. Sections across folds in Thessaly (*after Aubouin, 1959*)
   A thin sedimentary cover, ranging in age from Lias to Eocene, and topped by flysch, has been greatly folded and fractured. The main structural forms are homoclinal. The relief is on the whole moderate and is only a pale reflection of the deformation

1. Ionian flysch	4. Vigo limestones
2. Eocene limestones	5. Upper Lias and Dogger
3. Breccia with rudistes	6. Lower and Upper Lias

Following morphogenetic fashioning, the relief is more or less attenuated, the topographical gradients becoming progressively gentler than the structural gradients so that evolution in the residual phase is proportionately prolonged, that is, after the cessation of tectogenesis.

Box folds are particularly characteristic of rigid cover tectonics. Such folds are characterised by flat-bottomed synclines and anticlines with subhorizontal summits separated by severely inclined limbs. The cover imperfectly masks a structure of basement wedges. Typically, the box folds are of little extent and pass readily into domes and basins. The plateau of St Christol in Provence is an excellent example of this. It is rare for the box folds or the domes and basins to be affected by faults, although these frequently control the activity of secondary elements in the interior of domes and basins. An example of this can be seen in the plateau of St Christol, which is made up of a wedge of Cretaceous (notably Urgonian) limestones truncated by a deformed erosion surface. Here faults have produced small collapsed blocks in which are preserved discordant formations of soft rock, principally the molasse. Differential erosion has removed them little by little and depressions have appeared in their place (see geological map *Forcalquier*, scale 1:80 000).

The limbs of box folds are typically fragile. Their rigid strata have been subjected to tensional effects which have generally fragmented them. They are very fissured and thus susceptible to the development of cavernous hollows. The very creation of the folds which sometimes accompany earth-quakes facilitates their downwearing by landslides and scree development. Thus, the limbs of box folds are almost always dismembered and the synclinal depression at their foot is etched out on their margin either in the form of hills or as a glacis truncating the subvertical strata. Due to the climate, this is very common in North Africa where this style is widely distributed around certain rigid elements such as the High Plateaus. When lithological conditions are favourable, *barres*[1] and chevron folds follow as the anticlinal axis is approached.

Box folds pass into horsts in a whole series of transitions. In the most up-lifted part, there is frequently a horst which passes towards its extremities into a box where the beds, being less stretched, are not faulted. The boxed appearance is often attenuated in periclinal regions. These conditions favour the enhancement of lithological contrasts and, thanks to lower dips, annular valleys appear in the sectors instead of the *barres* of the culmination zone. These valleys are simply subsequent depressions below the crests and confined on their outer side by a resistant stratum.

Diapirs and mushroom folds

In contrast, *diapirs* are due to a hyperplasticity in the rock layers. They form within the cover rocks where beds of clay and plastic marls contain sufficiently extensive, thick lenses of evaporites, marine salt, gypsum, etc. These evaporites influence the tectonics in three ways all of which occur in varying degrees of importance:

1. They modify the consistency of the clays and yield alkaline waters which flocculate them and hence render them liquid. In this way the 'plasticity' of the clays grows. Their mobility is increased. To some degree they lubricate the tectonic processes. *Décollements*, distentions and com-pressions (*bourrages*) are greatly facilitated. All that is needed is that the evaporite crystals should mix with the clays, especially crystals of marine salt. The degree of displacement of the cover in the Jura and certain parts of Provence results from the presence of Triassic salt beds.

2. They constitute masses which behave plastically. In fact these evaporites dissolve and recrystallise readily which allows them to deform with ease. Also, they readily migrate as a result of regional tectonic stress or of isostatic mechanisms. The evaporite lenses have rarely preserved their initial forms. Very often, they have been heaped up one on another due to lateral com-pression so that they have become abnormally thick at the expense of their lateral extent.

3. Being less dense than the other rocks, they result in the release of any local isostatic disequilibrium. Due to their lower density, the bodies of salt have tended to migrate towards the surface every time it is not blocked by a

---

[1] Strike ridges made up of resistant members with very high, almost vertical dips—*Editor*.

sufficiently rigid caprock. Such movement is greatly facilitated when they are covered, as is so often the case, by clays which lose their consistency when in contact with salt water. They then up-dome the superjacent beds, fracturing the coherent horizons, and end up by arriving at the surface. This is a diapir or piercement structure. Evaporites do not necessarily produce diapirs. They do so in Alsace (Dôme de Hettenschlag), but neither in Lorraine nor in the vicinity of Paris. For the process to work, the salt beds must be sufficiently thick and, furthermore, be subjected to compression which allows them to thicken and to pile up one on another. Only then can isostatic uplift be produced. That is why diapirs are always associated with cover fold tectonics. They are not found in rigid regions made up of tabular beds when these are intensely faulted and when transverse faults release, as in Alsace, compression effects during the collapse of a graben. Northern Germany is a region of diapirs: again they occur in the folded cover. Algeria, Iran, the Corbières and around Sospel in the Niçois Alps, are also diapiric areas, again associated with folded cover rocks.

Diapirs produce a particular fold style. The effects of initial compression are effaced by the dislocation of the beds due to pressure from below produced by the diapiric plug in the course of its rise. Under such pressure the discrete rock layers tend to be pushed together. The mushroom fold or

FIG. 2.36. The flank of a diapir, at Sodom, in Israel, on the western edge of the Dead Sea

Beds of marl with thin layers of limestone, gypsum and salt have been given a vertical dip and in places are slightly overturned (nearest the camera). In the distance, at the end of the periclinal structure, the dips decrease. Under a desert climate, the structure is finely etched by stream action and each harder bed displays a *barre* cut into chevrons. At the foot of the hills the beds are mantled by *glacis* sediments

overhang (*pli en champignon*) is produced. The complete feature corresponds to the diapiric mass. Beneath the head of the mushroom, the beds become spread out and dislocated in the form of a fan. Above they still present their original continuity, becoming progressively more broken up at depth as the diapiric plug is approached.

When the diapiric core does not quite crop out, diapirs yield equi-dimensional domes, whose distended beds are rapidly denuded. Thus, the core is rapidly revealed and brought into relief by the erosion of the surrounding strata. In humid climates, salt domes revealed in this way are reduced by solution. The morphology they give rise to is never important and is limited to a minor doming of the soil surface. Above all it is at depth that the structures are deranged. Despite this, there is often deformation of terraces, e.g. at Hettenschlag (Rhine terraces). In deserts and arid areas, solution does not reduce the salt plug at the rate at which it is revealed and an extrusion of salt or gypsum can be produced at the surface. In Iran, evaporites flow in plastic fashion away from the domes in the manner of glaciers or mud flows (*glaciers de sel*). In such cases, diapiric relief is very positive.

It often happens that diapirs are emplaced during tectogenesis, as a result of compression. Masses of mobile salts, insinuating themselves into the zones of lower resistance, profit by *décollement* and, because they favour sliding, become laminated. They take on a generally starshaped appear-

FIG. 2.37. Folds and overfolds at La Roche des Arnaulds (Hautes-Alpes)
Jurassic (Bathonian) limestones make up this mountain. The general structure is anticlinal, but the relief only partly corresponds to the structure and does not reflect the overfolds. Locally (centre of picture) the flank of the mountain is made of a thick limestone bed that does give structural relief

ance. The structure becomes an extremely complex puzzle but, as a general rule, dislocation of the strata on the roof of the diapir facilitates dissection and usually produces an inversion of relief every time the diapiric core lies at not too great a depth. The environs of Sospel provide a good example (see geological map, *Nice* area, scale 1 : 80 000).

Conclusion

From geosynclines to *plis de couverture* we have phenomena at divergent scales: these phenomena get smaller and smaller. The geosyncline is measured in hundreds, even thousands of kilometres, the fold in single kilometres. The mechanisms envisaged vary according to the scale factor. Geosynclines involve the very structure of the crust of the earth itself, of isostatic compensation between sial and sima. The understanding of the architecture of a fold demands an analysis of how the diverse stresses and strains of regional earth movements react on a heterogeneous milieu: a heterogeneity made up not only of facies variations and structures affecting them (faults and prior folds), but also the existing relief.

Because of the nature of geomorphological phenomena, there is perennial interaction between the tectonic mechanisms and the processes of morphogenetic evolution. The unloading of cordilleras by dissection and the loading of geosynclinal troughs by sedimentation are geomorphic phenomena. They influence the tectonic mechanisms concurrently through the action of isostasy and the mechanical differences that this introduces. An accumulation of uncompacted muds recently laid down by turbidity currents does not react to the tectonic stimuli like ancient consolidated sediments or, for obvious reasons, as would a granitic basement. Now, if the emplacement of sediments is guided by subsidence, that is by tectogenesis, their nature is largely influenced by climate which controls the morphogenetic system. While less obvious in morphotectonics, perennial interaction between internal and external forces is no less important. The study of this interaction requires systematic team work between geomorphologists and geologists, although the point of view of the two related disciplines is different. It is by carefully placing all geomorphological knowledge into its geological framework that one can analyse it and understand it. To reconstruct the evolution of the earth's surface, it is essential to invoke geology, and if necessary, borrow its methods. In a subject which is complex enough, theoretical models which are the product of the imagination can only be delusive simplifications which pose a very great danger: to accept caricatures which, like the Davisian schemes, do not even resemble reality.

3
The platform regions

The platform regions make up the core of the continental blocks. As we have seen, geosynclines mould themselves around the borders of these cores or form in clefts in the surface of the continents. Some geologists accept a slow growth of the latter arising from progressive stabilisation of successive unilateral geosynclines. This phenomenon seems to be operative in certain cases, but it is not alone and several processes, antagonistic to it, act in the opposite direction. The stabilisation of geosynclines and their transformation into part of the platform is thwarted by the great rifts which cut into the structural part of certain continental blocks, like the Rift Valley system of East Africa, or like those which break up and transform parts of the continents in the development of regenerated geosynclines. In fact, some kind of balanced view, based on systematic palaeogeographical reconstructions, is needed. It has not yet been attempted.

Despite everything, one fact of observation is noteworthy. The continental platforms are predominantly at the surface of the globe and make up more than three-quarters of the present dry land area. Therefore, they provide the most widespread type of structure to be found in the continental masses. There are two other characteristics of equal import: the absence of seismicity outside the great rifts and the presence everywhere at relatively great depth of folded rocks resulting from geosynclinal development.

The shield part of the platforms is also made up of stabilised geosynclinal structures of great age which control the tectodynamic style. But this shield does not crop out everywhere. For it to be at the surface it must have resisted subsidence and even suffered some degree of uplift due to relatively permanent positive movements. Elsewhere, it has been overspread by detrital sediments, as in the inner delta of the Niger, or covered by epicontinental seas as in the present North Sea or the English Channel. It is then masked by sediments which in certain sedimentary basins, such as the Paris Basin, attain several thousands of metres in thickness. While the tectodynamics of both these uplifted and downwarped regions are of the same type, characterised by slow movements of great radius, the folds produced are very different and their contrasting geological evolution has resulted in different structural traits. It is these which control the contrast between sedimentary basins and ancient massifs in Hercynian Europe, which is the classic case.

We next examine the tectodynamics of platform regions, then the diverse structural types that they produce, with the corresponding relief forms, and finally the more local forms, shield contacts and coastal relief.

The tectodynamics and structural evolution of platforms

There is no sharp break between geosynclinal evolution and further development in the platform. The dying geosyncline is characterised by a progressive stabilisation which constitutes the transition phase. Production of folds, the typical architecture of geosynclinal development, is completed well before this stabilisation, even during the course of the orogenic phase: it is completed, it seems, in the Central Alps. During the Hercynian orogenesis of Europe, for example, a last folding spasm took place towards the end of the Carboniferous. In the Permian for some 20 to 25 million years this no longer prevailed, only fault block tectodynamics with localised accumulations in the basins, like those of Lodève or Saint-Dié, or increments of detrital deposits were maintained by the persistence of a slow uplift of neighbouring blocks sometimes accompanied by faulting and volcanic eruptions. Then, generally speaking, all this activity gradually dies out and the situation stabilises. Subsidence of the basins slows down and downwearing gradually reduces part of the residual relief. In the Trias, for example, marine sedimentation transgressed as far as Germany and, in the northeast of France, it forms a vast littoral plain covered by detrital sediments not so very different from those of the Permian but forming very extensive beds of very much greater regularity. Block movements occurring toward the end of this evolutionary sequence take the form of warping of very great curvature. Stabilisation is achieved and the passage from dying geosyncline to continental platform is realised.

These broad warping movements are characteristic of the tectodynamics of platforms; accordingly, we study them more closely.

Characteristics of platform tectodynamics

The stability of platforms has already been stressed. It is the basis of the arguments put forward by the old guard in the defence of the theories of W. M. Davis. For example, Baulig has supposed that the Massif Central has been stable since the Miocene and that, for the 28 million years which have elapsed since the beginning of that period, the successive episodes in its morphogenesis have been due solely to variations in sea level, to eustatism. Such a stability for a fragment of platform violently fractured on the edge of the Alpine geosyncline would be very surprising, especially during a period of vulcanism. But the very character of the movements which affect platforms renders their study delicate. Such study must be based on very precise paleogeographical reconstruction, the techniques of which are still being developed.

THE STYLE OF THE DEFORMATIONS

Platforms are characterised by the predominance of openfold structure, particularly well seen in areas of sedimentary cover. But as soon as the deformation becomes more intense fractures are also met with.

Openfold deformation of broad radius

This is expressed in the very low dips of the sedimentary rocks. While, in folded chains, the mean dips of structural units go from 20 to 60 degrees in the sedimentary basins, here they remain below 5 degrees. Over large parts of the Paris Basin, dips are around this order. They maintain this value in large areas of Paleozoic cover in the Sudan, in the interior of Brazil, and on the Mesozoic beds of the Russian platform. Dips higher than 5 degrees are generally localised in particular zones where they are associated with faults, as on the eastern border of the Bourguignonne scarp or in the Vosges. Frequency distribution curves show a very clear drop in the occurrence of dips with values around 5 degrees. A minor peak can be observed round about 10 degrees, corresponding to platform regions dislocated by faulting, then finally the important maximum occurs between 30 and 50 degrees which characterises the fold regions.

These low dips imply open folds of very large radius of curvature, for they vary very little in comparison with the much higher dips of the fold belts. They can be followed over considerable distances—scores of kilometres. From the edge of the Vosges to the neighbourhood of Paris, the diverse Mesozoic beds are always inclined in the same direction, that is to the west. The limestones which make up the Barrois plateaux at an altitude approaching 300 m are found around Paris at 400 to 500 m below sea level. Thus the reduction in level is about 800 m in little over 200 km, being a mean dip of a little more than a quarter of a degree.

It is more difficult to recognise similar deformation in the shield area. However, on their margin one can measure the inclination of ancient exhumed erosion surfaces like the post-Hercynian surface surrounding our ancient massifs. The results obtained are of the same order of magnitude: almost always less than 3 or 4 degrees and usually about 1 degree. Now these discordant surfaces are terrestrial erosion surfaces retouched by a marine transgression. Even where they are very regular, even planar, they must have had an initial slope, otherwise they would not have undergone a gradual transgression. In many regions, residual relief stands above the sediments slowly burying it, e.g. the quartzite *barres* of Falaise, Normandy. As is shown by the slightness of the measured slopes, the initial gradients of the discordant surfaces, while modest, were not negligible: the values obtained are little greater than the dips of the sedimentary beds of the basins. Moreover, where these surfaces are readily recognisable, the margins of such truncated shields coincide with rather more marked tectonic deformation and this has strictly controlled the edge of the sedimentary cover.

Sometimes undoubted traces of ancient erosion surfaces may be found, characterised by modification or by specific deposits, in the interior of shield regions. They have never been tilted to any great degree and the average calculable gradients remain slight, even without deducting those ultimately due to the action of faults.

Broadly warped structures give rise to an alternation of uplifted and downwarped units within the platform areas. They cannot be designated in the same terms as those of geosynclines because of the difference in their tectodynamic nature. The broad upswellings are termed *antéclises*, a handy term which allows the grouping together of the ancient massifs, the shields and the cratonic arches (*dorsales*). An antéclise always measures some thousands of square kilometres, sometimes several millions. The Ardennes, and its continuation the schistose Rhenish massif, or the Massif Central are antéclises, the same term encompassing the much vaster Scandinavian shield or the Guinean arch. Conversely, the elements subjected to downwarping in the same style are broad basins termed *synéclises*. The difference

FIG. 3.1. Folds and faults on the western border of the Dead Sea rift valley
    Alternations of marls and Cretaceous limestones that are differentially eroded in a semi-arid climate. On the left, tabular compartments delimited by faults. On the right and in the foreground, a slope corresponding to the bending of the strata

between a synéclise and a syncline is the same as that between an antéclise and an anticline. The Paris Basin and the Congo Basin, for example, are synéclises.

Antéclises and synéclises are regional units controlled by broad scale deformation. They are characterised by a general tendency to uplift (positive tectonics) or to subsidence (negative tectonics) which controls, as we shall see later on, all their geological evolution and their structural and morphogenetic characteristics. But this general determining tendency does not rule out more varied deformations of smaller dimensions, sometimes working in the opposite direction. Thus, in the Paris Basin synéclise on the edge of the English Channel, undulations appear giving rise to anticlinal upswellings and synclinal gutters: Boulonnais and Bray on one side, and the trenches of the Somme and the Seine on the other. But here it is a matter of subordinate features, minor in relation to the general feature.

In regions of low, consistent (homoclinal) dips, so extensive in the northeast of the Paris Basin for example, the dips of the beds are not always constant even in the absence of fractures. They often present local variations which, on their own, only a very detailed study can evaluate. We have been able to show, also, an alternation of *flexures* where the dip increases and levels (*paliers*) where they diminish. Around Nancy where the mean dip is about 1 to 2 degrees, it can reach 3 to 4 degrees on the flexures, becoming imperceptible on the levels. These variations in dip influence the dissection of cuesta landscapes to a considerable degree. Despite their smallness geomorphologists should not neglect them, for most geologists have scarcely accorded them much importance. Large-scale deformations are also accompanied by a variety of subordinate features.

Minor features

The rigidity of the granitic shields, much the commonest type, give a brittle look to their tectonic style. Fractures are numerous in the granite–gneiss series and clearly more rare in the poorly metamorphosed schists which, being more supple, undulate more readily without yielding to fracture. As soon as the rigid shields are subjected to somewhat intense or sharp stresses they yield by fracture. It is these which yield the finest crustal rifts such as the Rift Valley system of East Africa–Red Sea–Palestine. On the margins of the geosynclines, the platforms often break up, as in the Vosges–Black Forest ensemble or in the Massif Central with the Limagne and the Forez. Here it is not simply a case of minor features: very much to the contrary. By their importance and the size of the compartments that they delimit, they are the equivalents of the geosynclinal wedges. These are considered in Chapter 4.

In contrast, the platforms are affected by numerous structural features of more limited extent, such as fractures, among others. It is easy to spot them

FIG. 3.2. The flexure that bounds the western side of the Dead Sea trench, in the hills of Tamar
    The beds of limestone and marl are strongly bent, and a resistant bed outlines a structural surface

in the sedimentary cover. For example, in the Paris Basin, the faults generally have a feeble throw varying from a few metres to one or two hundreds at the most, but typically extending for great distances, especially those which have a fairly large throw. This is the case, for example, in the Bray fault which extends for more than 100 km, or the Basse–Seine fault which runs parallel to it. In the Nivernais the faults of the Auvergne continue to the northwest of the Morvan but rapidly lose their importance further north: at the same time, they become attenuated and less dense before they disappear entirely (see Geological map, *Clamecy*, 1:80 000 scale). In the Haute–Marne, a narrow fault trough can be followed for some kilometres.

On thick sedimentary covers, such as those of the Paris Basin, where plastic marly beds are well represented, a veritable cushion is produced. The faults die out upwards. They are more numerous and more important in the lower part of the cover close to the rigid shield and rarer and with a smaller throw in the upper part of the cover where only the most vigorous faults make themselves felt and, even then, have their throw reduced. Numerous faults, generally with a weak throw (only a few metres), are visible, for example, around Nancy. In the chalk of Champagne, where some tensional stress has occurred, there are practically none and, being damped down, their effect has been limited to a fragmentation of the fragile rocks by dense fissures oriented parallel to the structural features of the neighbouring regions.

Certain fractures in the sedimentary cover are accompanied by light tangential movements which cause undulations to appear. This is true of the northwestern part of the Paris Basin, in Boulonnais and in Perche. Some faults are associated with the synclinal trench of the Seine and the up-archings of Bray and Boulonnais. Some periods of compression seem to have provoked crushing along these fractures, while periods of tension, which alternated with them, have allowed the downwarping of strips whose dips are abnormally high in relation to those of neighbouring regions, sometimes reaching 10 degrees or so. They are tied to local tilted blocks and are always clearly circumscribed.

However, the undulations on the northwest of the Paris Basin are an attenuated variety of this genre of structural features. Normally, they are more vigorous: in fact, we pass by imperceptible stages from undulations of this type to *plis de couverture* as sharp folds become accentuated along the more important fractures or where the horizontal displacements have been greater. This can be seen in the southeast of the London Basin where dips of 5 to 10 degrees are not exceptional on the flanks of folds, and also in the Weser Hills where the cover folding is clearly outlined. The association of these open folds with faults is explained by moderate *décollement* acting on flexible cover rocks, probably accompanied by phenomena arising from anchorage of the material on the faults of the subjacent shield. We thus pass gradually into the comtois style.

In contrast, the folds are less marked in the cover rocks which have not suffered tangential displacements. This is the case in Lorraine where, by the end of the nineteenth century, geologists had established anticlinal and synclinal axes in Jurassic beds with a vertical range of some tens of metres and a horizontal wavelength of some tens of kilometres. They are linked to the levels and flexures already described. For example, the synclinal axis passes by the town of Toul, an anticlinal axis by Pont-à-Mousson and a new syncline towards Metz. Orientated northeast to southwest, they prolong the obviously more marked undulations which affect the Palaeozoic of the Sarre coalfield. Some Hercynian structures, not rendered rigid by metamorphism, have guided the later deformations, folds of slight dip associated with numerous fractures generally of small size. Here we are dealing with posthumous movements.

The cover rocks can thus present a variety of structural patterns depending on the degree of thrust, the effect of compression, and deformations of the substratum, which themselves vary according to whether they rest on a more rigid and broken granite gneiss shield or on less indurated and more supple folded sedimentary formations. The rift valley systems, accompanied by tangential displacements with alternations of phases of relaxation and compression, give rise to more severe upswellings and downwarped strips which pass very gradually into cover rock folds.

In the platform regions, the critical role is played by the shield because of the high degree of independence of the cover rocks. When *décollement* really begins we have typical cover rock folding. This gives rise particularly to cushioning effects (*matelassage*). One must expect to encounter in the shields themselves, therefore, a more vigorous tectonism than in the cover rocks, whatever the tectonic style. Because of the greater rigidity of the shield, faults play a major role in this distinction. Except for particular cases, this has been misunderstood, even systematically ignored, for a very long time, because it conflicted with the erosion-surface mentality based on the twin abuse of the surveying aneroid and graph paper! Every marked slope on the shields was interpreted automatically as the link between two cyclic erosion surfaces, a facile practice which still rages in some countries. In fact, petrographic studies often show differences in rock type from one part of these surfaces to another so that these turn out to be variously degraded fault scarps.

This tectonic style, characterised by the predominance of movements with a big radius of curvature, and of moderate warping interrupted here and there by fractures, is typical of platform regions. Furthermore, it accords with some distinctive tectodynamic rhythms.

THE RHYTHM OF THE DEFORMATIONS

Platforms are sometimes subjected to extremely rapid movements, of a speed equal to those of the geosynclinal cordilleras. This is the case with the postglacial isostatic uplift in Scandinavia, which is as rapid as the present day earth movements in New Zealand. But in this case it is a question of a particular mechanism, of an exceptionally vigorous isostatic reaction due to major disequilibrium produced by glaciation followed by rapid deglaciation. The two things are not, in fact, comparable. However, it does help to bring out the great susceptibility of shields to isostatic mechanisms.

Usually, the rhythm of deformation is slow on platforms. Some are characterised by a remarkable stability. Madame Ters, for example, has been able to show in Vendée surfaces of Cenomanian, Tertiary and Quaternary age one inside the other and almost at the same altitude. That does not rule out intermediate pulsations, but these must have been of a very limited amplitude otherwise the traces of ancient levels would have been effaced between times. Furthermore, these pulsations are produced round about a mean position which scarcely changes. But that does not rule out the subsidence in the Pliocene of a structural compartment (that of the Lac de Grandlieu) a short distance away. One might object, and with reason, that Vendée constitutes part of the Armorican Massif where the positive tectonic tendency has never been very strong and which is, because of this, an antéclise of little vigour. However, even in some more pronounced massifs such as the Vosges or the Ardennes, uplift remains moderate. The Hohneck and the highest summits of the Vosges do not lie very much below the level of the post-Hercynian surface, for one finds in this area remains of pre-Triassic alteration in the joints and it is known that such alteration does not penetrate beyond 200 to 300 m below the surface. On the highest summits of the Ardennes are found reworked remains of flints derived from the chalk. It follows that the high points of the shield have not suffered a very considerable downwearing. It is a long way from the 400 m average downwearing which was produced in the Italian Alps in the Quaternary alone.

For the same reason, sedimentation in the synéclises is a long time in reaching the same thicknesses as those in the geosynclinal trenches and troughs. Typically, it does not exceed two to three thousand metres as is shown in the Hercynian basins of Europe. These accumulations form over extremely long time periods. In the Paris Basin, for example, between 2 000 and 3 000 m of beds were deposited between the beginning of the Liassic and the middle Tertiary, i.e. some 120 million years. Since the Miocene (less than 15 million years), the same thickness has been deposited in the plain of the Po which is a geosynclinal trench. This gives a general idea of the order of magnitude involved. As described below (p. 183), it results in some very important differences in the very nature of the sediments and in their disposition.

In comparison to geosynclines, then, platforms are regions of slow-acting tectodynamics, where movements last a long time and are less violent. However, that does not prevent the appearance, in certain epochs, of severe paroxysms which originate in wideranging events which also typically affect the neighbouring geosynclines. This is true of the Oligocene faults of the Massif Central, of the Vosges and of Central Germany: all are the results of very violent stresses which have affected the European continental surface as well as the Alpine geosyncline. Apart from such paroxysms, deformation is steady and gradual. Deformation is not associated with severe breaks in equilibrium like those of the geosynclines, so that it is not expressed by earthquakes. In contrast to geosynclines, platforms are not seismic regions, apart from the great active rift valley systems.

This typically lengthy process of deformation and the gradual character of movements in antéclises and synéclises does not mean that the tectonic trends remain constant. On the contrary, and as in the geosynclines, they are subjected from time to time to reversals. In certain periods, the subsidence of the sedimentary basins is checked, and uplift of the antéclises gives place to an approximate stability, even to a degree of subsidence.

Take the Paris Basin for example. This appeared only at the end of the Triassic. In fact, sedimentation had been important in Lorraine during the Triassic only in a gulf opening towards the north-east in the direction of Hesse. By the Liassic, in contrast, the slow downwarping allowed a broad marine transgression on to the margins of the Massif Central which, in Normandy, was reunited with the epicontinental sea occupying the London Basin. The subsidence continued for some 25 million years, with only minor modifications, until the end of the Jurassic. Thus, in the Aalenian, some warps developed in Lorraine, with reactivation of the Hercynian anticlines into posthumous folds. Gradually during the Jurassic, the sea advanced to the borders of the Ardennes in a northwesterly direction, that is obliquely with reference to the actual border of the massif, which nevertheless remained dry land. At the end of the Jurassic, the cessation of subsidence produced a general emergence. An intermittent gulf maintained itself after a fashion in Burgundy, opening towards the seas bordering the Alpine geosyncline. But the Vosges and the Ardennes were uplifted and this warped the strata. Once emerged, the Jurassic land areas were attacked by erosion. Indeed, karst formed at the foot of the Ardennes. Gradually, the surface of erosion became established which, modified by marine transgression in the middle of the Cretaceous, became a discordant surface. The same thing happened in the south of England, where the folding appeared during this emergence. It lasted a good 15 million years. After that, subsidence resumed throughout all of the upper Cretaceous (35–40 million years), interrupted by a new widespread emergence at the end of this period, in the Danian. However, during this renewed period of sedimentation, the palaeogeography changed. The sea in the Paris Basin no longer seems to have opened towards the Alpine geosyncline, but rather to the north and northwest. The up-arched area of Artois, which had previously limited it towards the north, became submerged. There the chalk rests directly on the Palaeozoic in an extensive disconformity.

Throughout the 30 million years that Eocene and Oligocene sedimentation persisted, subsidence was slower and more irregular, and affected a more limited area. That is why lacustrine and marine formations alternate in a complicated fashion. This is the prelude to the emergence which began at the end of the Oligocene, 28 million years ago.

Palaeogeographical reconstructions allow us to say that after these events the antéclises around the Paris Basin were themselves uplifted in varying degrees. For example the Ardennes, which had resisted submergence since the end of the Palaeozoic, ended by being invaded by the upper Cretaceous Sea at the same time as the up-arching of Artois. It appears that at this point in time, the positive movements which had characterised it until now were interrupted, after more than 90 million years. Until the Miocene, this tendency lay dormant and the Ardennes seem to have retained (as is still the case in the Artesian arch) their Cretaceous cover, to which have been added, sporadically, thin Tertiary beds. The duration of this phase was about 75 million years. Towards the end of the Miocene, positive movements were again renewed, isolating the Ardennes which became further uplifted from the Artesian arch which was raised much less. The cover rock was then stripped and the shield dissected. That lasted from 15 to 20 million years.

Many examples are to be found throughout the Hercynian zone of Europe from the London Basin to the Bohemian massif. They show pulses and reversals of tectonic trend such as occur in the cordilleran phase of the geosynclines. But their rhythm is much slower and each phase lasts from five to ten times as long. That is why disturbances of the equilibrium are less marked and one may speak of a relative stability. Readjustments, notably those of isostatic type, have time to operate at the scale of the vast structural units characteristic of the platforms. As in the case of geosynclinal troughs, but to a much smaller degree, it would seem that subsidence of the synéclises was limited which would explain the periodic interruptions in downwarping and to a certain extent its checking such that the sedimentary basin was made to emerge, so allowing its dissection. But on the edge of the emerged sedimentary basins there is much that is still submerged and which forms a part of the epicontinental seas. This is the case in the English Channel.

These slow pulsations of the earth's crust, interrupted from time to time by more violent movements often accompanied by faulting, give a rhythm to the geomorphological development of the platforms.

General geomorphological evolution of the platforms

Positive tectonic movements cause erosion to predominate in the antéclises and this erosion provides the sedimentary materials characteristic of subsiding regions. The sedimentary series of the synéclises together constitute the correlative formations which allow the reconstruction of the geomorphological evolution of the antéclises. In this there is an important point to examine. We shall deal with it first, and then we shall study how the different types of tectonic movement underlie the origin of various types of platforms.

EROSION AND SEDIMENTATION ON THE PLATFORMS

We have seen that deformation of the platforms is generally slower and more continuous than that of the geosynclines. In the case of antéclises, dissection has more time to become organised. Furthermore, the predominance of broad radius movements gives to this dissection a more gradual appearance than the attack on the edge of a rapidly uplifted wedge. Finally, the lithology also plays a part. These different factors influence the manner of downwasting and, as a result, the nature and rhythm of the resulting sediments.

Lithology of the antéclises

In a general way, the antéclises and synéclises of the platforms are a long way from coinciding with the structural units characteristic of the end phases of the geosynclinal sequence. Thus, the centre of the Paris Basin is characterised by a subsurface of granitised metamorphic rocks. It probably corresponds to a region of the Hercynian chain which has been profoundly eroded and in which, therefore, uplift has been more intense. In contrast, the Ardennes and the Artois arch are entirely sedimentary, their cover folds and coalfield deposits indicating a marginal, less elevated region of the chain. In Lorraine, the Paris Basin extends to the Carboniferous folds whose outcrops are cut across by it.

Accordingly, actual antéclises on the platforms do not necessarily coincide with the most uplifted structural units in the course of the waning evolution of the geosyncline which preceded their appearance. This would explain why three different groups of rocks are met with: these are particularly apparent in the Armorican Massif which, probably due to its rather limited uplift, has suffered only a moderate degree of downwearing since the Palaeozoic:

1. *Sedimentary rocks*, ancient folded cover rocks, in which shales, sandstones and quartzites, and sometimes limestones are of major importance. This is true of the Ardennes, the Appalachians in the United States, and a major part of the Armorican Massif. These ancient rocks are always folded, by definition, and cut across by surfaces of disconformity on which have been deposited platform formations lying in tabular fashion. In regions which have been prematurely consolidated into platforms, the cover rocks may often, therefore, be ancient Palaeozoic, even Eocambrian rocks, i.e. more ancient than the beds which make up parts of other, more recent shields and which are themselves folded. In the coalfield basins of western Europe, the Carboniferous rocks have been folded by the Hercynian orogenesis. In contrast, the Devonian of the Scandinavian Shield is a tabular cover in the island of Åland. In the same way, Ordovician rocks form vestiges of a tabular cover on the Laurentian Shield. It is the *relative* age of the beds, controlling their place in the development, which really counts, not their absolute age. The folds which affect the sedimentary beds incorporated into the shields are of various styles. As well as the broad regular folds of 'Jurassic' style, like those of the Appalachians, overthrusts occur (e.g. northern coal basin of France and the Pas-de-Calais). While becoming indurated in the course of time (for example clays are always transformed into shales), the folded sedimentary beds of the shields present an interesting geomorphological situation above all when they are made up of ancient accumulations of epicontinental seas: they provide conditions which favour differential erosion to produce the *Appalachian relief* type.

2. *Metamorphic rocks*, products of the modification in depth of what were initially sedimentary rocks, due to the effect of temperature, pressure and the proximity of igneous magma. Due to the persistence of downwarping, they are gradually modified to adapt to a physico-chemical milieu completely different from that in which they were laid down. In *regional metamorphism*, there are thus crescentic zones of metamorphism. With pelitic sediments, one passes at increasing depth from a sedimentary shale to a lightly metamorphosed schist, for example a *schiste lustré*, and then to the mica schists—first those which are less rich in metamorphic minerals (low grade mica schists) and then to the richer ones (high grade mica schists). Thereafter, one passes to gneiss and then to granites. Generally

speaking, the granites take the form of enormous masses which have digested the sedimentary rocks to make up vast flattened domes. Their density being only about 2·65, some authors have favoured mechanisms similar to diapirism which would explain the very long, narrow anticlinal (*brachyanticline*) appearance of these granitic masses. They would, in this case, play a role in the positive movements of antéclises. While this may be so, when these regional metamorphic structures are attacked by erosion, the resulting topographic surface often cuts across the various metamorphic zones which form aureoles surrounding the granitic massif. For this reason we sometimes designate this gradation of metamorphism as *zoneography*. A fairly gradual transition is not so favourable to the development of a landscape of differential denudation, although the style of dissection differs in mica schists and in granites.

In contrast there is another type of mechanism which gives rise to some extent to the differentiation of lithological relief elements: *contact metamorphism*. In this case, a granitic mass has cut across other rocks in the process of digesting them in the same way as the ascent of lavas in the volcanic regions. This granitic mass forms a batholith The great difference is that in the case of volcanism the lavas normally reach the surface and this is never the case with a batholith. An important modification of the rocks surrounding the batholith occurs, known as contact metamorphism. Being rapidly recrystallised, these rocks are formed of very small crystals which, where they are in contact with the granite, give them a horny appearance and a glossy fracture, whence the general name *hornfels* is derived, a term borrowed directly from the German. Usually, hornfelsen are very tough and often brought into relief by differential erosion of the surrounding formations as much as of the batholith itself. Usually, they form only a narrow aureole (often cut by veins which are generally metalliferous) of some hundreds of metres to, at most, a few kilometres in breadth. Batholiths reach some tens of kilometres in diameter. They are not always granitic but are made up also of diorites and syenites.

3. *Igneous rocks*, products of magma which has consumed metamorphic and sedimentary rocks but has left them unrecognisable. In the case of gneiss, the most evolved type in the metamorphic series, the disposition of crystals remains influenced by the laminations of the original sedimentary rock. It is not the same in the igneous rocks whose crystals are disordered. The classification of these is a function of the nature of the crystals and the minerals which form them. That is why we speak sometimes of *crystalline rocks*. Granite is the principal member of a series containing many rocks which are related to it: granulite, diorite, granodiorite, syenite etc. All traces of stratification planes having disappeared, joints play an important part in the atmospheric modification of these massive rocks, because they permit the penetration of water which also controls breakdown by frost shatter. Some of these joints are the result of compressive and

Fig. 3.3. Granitic intrusions of
Taourirts in the western Ahaggar (*after
Bois-Sonnas, 1963*)
1. Granitic intrusions, with indications
    of annular structure
2. More recent intrusions
3. Stable block of the central
    Ahaggar, bounded on the west by a
    sharp break in relief, running
    north–south
4. Recent volcanic massif of Tahalra
5. Motor roads
Left blank are the Pharusian strata into
which the granitic rocks were intruded.
Such intrusions take the form of domes,
etched out by differential erosion

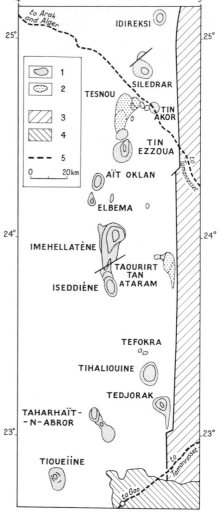

tensional stresses arising from the original tectonic movement. Yet others
seem to be produced by the effects of unloading which result from the
removal of the superjacent rock cover. Variable jointing can also be a
cause of the development of differentially eroded relief. In western Africa
the most ancient Pre-Cambrian granites, which have been subjected to
many periods of earth movements, are very broken up, while the most
recent granites, which formed batholiths after the beginning of platform
development, are left as massifs. The recent granites thus give rise to
mountain masses dominating lower regions cut into the ancient granites.

On the whole, therefore, the distinctive lithological characteristics of
the *antéclises* are only moderately favourable to the development of a land-

scape of differential erosion. This is only truly accomplished in the Appalachian structures. It is sometimes fairly clear in areas of contact metamorphism. It is seen much less in areas of regional metamorphism where it gives place merely to differences in style of dissection as expressed in the density of the drainage net, slope profile and slope steepness.

Tectonic style and modes of dissection

The temporal predominance of relatively slow movements and broad radius folding influences the manner of dissection of uplifted platform regions in a way which is very different from areas of sharper and more localised movements which affect geosynclines, especially in the cordilleran phase.

It so happens that the hydrographic network conforms to the general tectonic situation. The radial appearance of the drainage net of the Limousin emphasises the very distinctive antéclise which affects that region. The asymmetrical network of the Cévennes, as clear as that in Liguria, expresses the tilting of the southeast of the Massif Central. However, in the latter case, this does not record the results of typical platform tectonism: on the contrary, it is the result of a recent and severe paroxysm which produced relative movement between blocks.

For the hydrographic network to emphasise the details of the folding, the latter has to have been either violent and severe (which brings us back to the geosynclinal wedges) or, on the contrary, slow but very prolonged and very constant. This is the case in the Limousin which, since at least the beginning of the Tertiary, has functioned as an antéclise and as a source of detrital sediments which have been distributed over the neighbouring regions of sedimentary cover. Since then, it is conceivable that the hydrographic network, at least in its trunk streams, has been able to adapt itself to the form of this antéclise. However, in detail it has been influenced just as much by block movements which have been revealed in the very centre of the antéclise.

Nevertheless, it often happens that the relationships between the drainage network and the structure are very complex. This is so in two principal cases: (*a*) in the case of a discordant cover; and (*b*) when there is a basic change in deformation with time. The Paris Basin and the Ardennes provide us with excellent examples of such developments.

In the northeast of the Paris Basin, the Meurthe and Moselle river systems do not at all correspond with the structure. Their upper branches, running off the uplifted and tilted block of the Vosges, do not follow the regional dip of the beds towards the west where they border the edge of the Basin. On the contrary, they run obliquely towards the northwest cutting across several aureoles. The Meuse, which turns away from the Basin so very close to the plains of the Saône, does likewise beneath an even stronger scarp which causes it sometimes to run almost at right angles to the dip.

The three rivers converge in the area of Toul, where the Meurthe and Moselle were confluent until the Riss glaciation. Since then they have continued their course and, running very obliquely across the dip of the beds, they leave the Basin to become entrenched into the Ardennes and the schistose massifs of the Rhineland. It has been possible to show by means of palaeogeographical reconstruction, that this situation was an inherited one, a survival from the Miocene, which has since been modified.

In fact, at the beginning of the Miocene, about 25–30 million years ago, the Ardennes had not yet been uplifted to their present level. Around their highest summits are found the remains of a Miocene shoreline with chalky flints derived from a freshly eroded cover. In contrast, the uplift of the Vosges is more ancient: it got under way in the lower Oligocene as is shown by the filling of the Alsatian graben, say about 35–40 million years ago. At the beginning of the Miocene, an erosion surface descending from the Vosges towards the North Sea truncated the Mesozoic beds in Lorraine and then in the Ardennes where the cover rock was more or less completely removed. It is on this surface that the rivers were initiated during the Helvetian period as a humid climate replaced the semi-arid climates of the Burdigalian. But the tectodynamics became modified in the Plio-Quaternary. Both the northeastern part of the Paris Basin and the Ardennes were uplifted, but in so doing they became warped towards the west. The Meuse and the Meurthe were incised into the Ardennes and the Rhenish Massif in superimposed valleys the cutting of which was sufficiently difficult to provoke a rearrangement of the network. The Moselle ceased to flow into the Meuse and became diverted, towards the end of the Riss, into the Meurthe. Several western affluents of the Meuse (the upper Aisne and the Aire) became diverted towards the Seine in a series of captures. Nevertheless their upper reaches, including that of the Marne, still remained oriented towards the northwest conforming with the slope of the Oligo-Miocene erosion surface.

Further readjustments of the river network took place in the Paris Basin during the Upper Tertiary. At the beginning of the Miocene, the Loire flowed towards Paris and joined the Seine whose trace it broadly followed. The train of granitic sands known as the Sables de Lozère, dating from the Burdigalian, mark this course through the lower Seine. Subsidence of the Sologne intervened to make this course more and more difficult, so that finally the Loire was diverted directly towards the ocean, a change of course which was facilitated by the Pliocene downwarping of the Faluns Sea which formed a gulf in the region of Nantes.

Although these cases differ, particularly from the structural point of view, from those we have put forward in relation to geosynclines, it is clear that Davisian concepts are no more appropriate here than for the latter. There exist at one and the same time antecedence on the slow broad warpings and superimposition starting on an erosion surface which cuts across beds whose dips diverge from it. Now this erosion surface was not buried under a discordant cover: it bore only thin, localised and very discontinuous deposits, for the most part formed by semi-arid sheetwash.

Similar mechanisms occur frequently in platform regions and they have been recognised notably in the Congo Basin and in the interior plateaux of Brazil. The very character of the tectonic movements of the platforms, slow and of long duration, leads to the development of antecedent gorges whenever the climate is favourable. Broad-scale warping does not give rise to a renewal of severe erosion on the antéclises so different from that produced on the margins of fault blocks. On the contrary, it produces a gradual incision. On varying lithology, this stimulates differential erosion. The water courses, which cut into unconsolidated and, above all, impermeable marls make more rapid progress in comparison with those which are incised into limestones or those which suffer subsurface abstraction as in a karst region. Those which follow the dip, especially if situated on the less uplifted blocks (where in becoming incised, they do not have to overcome the effects of uplift), are in an equally favoured situation. Particularly in periods of alluviation, the rivers are very sensitive to warping which affects the plains on which they sprawl and which so encourages their wanderings. This is what happened in the case of the Loire towards the end of the Burdigalian, and on a very much smaller scale at the end of the Riss around Val de L'Asne with the diversion of the Moselle toward the Meurthe.

Differential erosion, especially when it operates in a situation where downcutting is predominant, results in frequent rearrangement of the drainage network in sedimentary regions, their more varied rock assemblages being less resistant than those of the granite-gneiss shields. As we shall see, such drainage developments are important on Appalachian structures, especially in cover rock regions.

The nature of the deposits

While in the geosynclinal context tectonic movements, because of their intensity and sharpness, have a definite role to play in sedimentation, their influence on the latter in platform regions is much more modest. In fact, it is the climatic influence which predominates, often in a very marked fashion. Without anticipating too much some questions which are dealt with in detail later on, some general indications of the behaviour of granite–gneiss shields under the principal morphoclimatic regimes may be given.

In a hot humid climate (intertropical forests), the granitic rocks are readily susceptible to intense chemical alteration which provides abundant dissolved matter, clays and finely divided detrital quartz (predominantly silts). These are readily mobilised and are easily evacuated via the drainage basins to the sea. The sands, taken up by littoral drift, build up the beaches and coastal barriers. Mixed with the fine fractions, they are sometimes entrained and moved directly towards the abyssal plains along submarine canyons where the continental block is suddenly terminated by the huge monocline known as the *continental flexure*. This occurs at the mouth of the Congo. Elsewhere the clays settle out slowly, suffering alteration and in reaction with the dissolved salts in sea water ultimately constitute new minerals, the *neogenetic clays*. The dissolved products are in part fixed by living organisms. For example, the lime derived from alkaline felspars helps to feed the corals, and the dissolved silica to supply the organisms with a siliceous skeleton, e.g. the sponges. The sediments deposited in the epicontinental seas are fine: argillaceous sands and sandy clays, marls with clay and lime admixtures, and limestones more or less marly. A more vigorously dissected region provides abundant clay, turbid sediments being more plentiful and less readily intercepted in their course across the flood plain. An area in an advanced stage of landscape reduction produces, above all, products in solution: the organogenic facies, such as the limestones, predominate around its margins. We know, for example, that coral reefs develop better in clear waters.

In tropical climates with a dry season, chemical alteration is less marked. The proportion of sand, particularly medium sand, is greater in the water courses. The proportion of dissolved products and clays is less. In the rainy season the debris is readily transported to the sea. The sedimentation is not fundamentally different from that of the hot humid regions, but the part played by clastic sediments is greater. The sandy clays and clayey sands constitute a larger proportion of the sedimentary assemblage. Clayey particles and iron oxides coating the sands are abundant and often give a reddish colour to the clayey-sandy sediments, particularly around the shores where sedimentation is rapid. This is especially so in the sandstones of the 'Germanic' facies of the Trias.

In the dry climates granular disaggregation predominates, essentially producing sands with a little clay and some gravel, particularly of fragments of vein quartz. This process can be relatively rapid, but the wadis drop the greater part of their debris in the piedmont areas where they spread out in broad sheets. Also to be found are sandy gritstone formations interbedded with predominantly siliceous and poorly rounded pebbles. Sandy clay lenses are also met with and owe their origin to settling out in distal depressions. These alternations of sandy gritstone and sandy clay can make up great thicknesses in the endoreic basins, so frequent in the subsidence zones under these climates, for they prevent the inadequate surface flow from reaching the sea and catch it, as it were, in a trap. Such series, described vaguely as *continental deposits* by geologists (who interpret them more or less correctly), are common in platform regions and as widespread as the sediments of epicontinental seas. They are dominant, of course, on the great continental masses: in Central Asia, in Africa, in South America and Australia. They make up a very important lithological type of the synéclises.

As in pre-Quaternary geological history the temperate climates have played only a small part and the cold climates have appeared only sporadically, we shall draw the line there. The sedimentary differences corresponding to the main varieties of climates which have predominated in the past demonstrate that the influence on sedimentation of the palaeoclimatic oscillations has been very great indeed. During the humid periods with intense chemical action, organo-argillaceous sedimentation predominated in the epicontinental seas. During the drier periods, with chemical action diminishing, the foreign detrital elements were carried away with sandy clay materials whose fairly rapid accumulation served to bury the organogenic component in their mass. Within certain limits of climate, a severely dissected relief favours this type of sedimentation. Leaving aside vulcanism and major earthquake faulting, the tectonic movements of the platform, being slow pulsations, never have severe repercussions on dissection. The often rapid changes from a marly clay to a calcareous sedimentation or to sand deposition cannot, as too many geologists believe, be attributed to tectonism. Certain workers have gone to the ultimate extreme of this view by attributing to tectonic pulsation alternations of sands and clays of only a few centimetres thickness repeated dozens of times in certain stages. On the contrary, they are explained by alternating phases of sediment transport and settling out in basins or littoral zones during successive floods.

But the vegetation also plays a part, giving some resonance to the climatic oscillations. H. Ehrart has highlighted the importance of vegetation in his theory of biorhexistasy. A climatic change towards drier conditions affects the vegetation cover which becomes impoverished and therefore less successful in protecting the soil. The alteration products of the preceding period, which developed to quite a thickness in the case of warm, humid climates, are vigorously scoured. Clastic deposits are produced during the period of disequilibrium, that is, the rhexistasy. In contrast, a climatic shift toward greater humidity allows the development of a protective vegetal cover and a reduction in sediment transport. The alternations of marl–clay and limestone facies in the Jurassic of the Hercynian basins of western Europe is explained much more successfully by climatic oscillations and the action of biorhexistasy, than by hypothetical staccato earth movements which fly in the face of what we know about the rhythm of platform deformation.

STRUCTURAL DIFFERENTIATION OF THE PLATFORMS

Two major factors affect differences between platform regions and, acting together, account for their subdivisions: the duration of development of the platform; and the style of the tectonic disturbances.

The duration of development

Certain platforms date back to the upper Pre-Cambrian, the stabilisation of their component geosynclines taking place some 600 million years ago, or even a little earlier. Other platforms appeared only at the end of the Hercynian orogenesis, that is only 200 million years ago. While these age differences are rooted in the distant past, they are none the less important and form the basis of a structural differentiation of the platforms.

In a general way, the most ancient fragments of the platforms are also the most stable, having a rigidity which allows them to resist even major tectonic stresses. For example, the Colorado Plateau, granitised in the Pre-Cambrian, is not ruptured and has preserved a tabular cover, despite its incorporation in the circum-Pacific mountain chain (itself uplifted several thousands of metres). Similar structures seem to exist in Tibet and in the Altiplano of Bolivia. In contrast, the Hercynian platforms, being younger and less indurated, have been rather poorly resistant to recent tectonic stresses. As we have seen, they have been fairly readily transformed into regenerated geosynclines, or they have been fractured and their fragments incorporated into recent geosynclines like the central Alpine massifs.

Being resistant to folding, several fragments of ancient, long-indurated Pre-Cambrian platforms have acted as rigid cores during later periods. When they are affected by a moderate degree of positive tectonism, they constitute shields.

Let us take, for example, the Scandinavian Shield. Its core, running from Sweden under the Baltic into Finland, is made up of extremely rigid Pre-Cambrian granite–gneiss rocks which, however, are perfectly capable of suffering rapid deformation into a flattened dome resulting from glacio-isostatic readjustments. To the west, towards the edge of the continental mass, it is bordered by the Scandes where a cover of Cambrian and Silurian sediments was vigorously folded during the Caledonian orogenesis, about 330 million years ago. The folds have even overlapped the margin of the shield towards the Swedish–Norwegian frontier. This ancient, largely granitised geosynclinal chain is itself now stabilised and transformed into a platform. Interdependent with the shield, it is only affected by movements of great curvature which have uplifted it in broad swells and broken it on the margins of the continental block adjacent to the northeastern Atlantic Basin. On its southeastern side the Scandinavian Shield declines gently beneath a tabular cover of Cambrian and Silurian age, this platform sedimentation including shales and sandstones overlain by limestones. It appears around Leningrad, in Estonia, and in the Swedish islands of the Baltic. Despite its age, the dip of the beds is no greater than in the Paris Basin. A similar situation can be observed in the northeastern United States. The granitic shield of the Adirondacks is flanked by the ancient geosyncline of the Taconic orogenesis (contemporaneous with the Caledonian). A discordant, perfectly tabular cover of Ordovician shales and Devonian limestones overlies the shield as well as the geosyncline which is moulded about its border.

While the shields and the old geosynclines welded around them have had almost the same tectodynamic evolution during the history of the platform, they display some geomorphological differences because the geosynclines are incompletely granitised, while the shields are completely so. In these geosynclines, the sedimentary fill has generally suffered an incomplete regional metamorphism (zone of mica-schists), pierced by granitic intrusions. A degree of differential erosion with the outlining of Appalachian forms may therefore be produced.

The recent platforms, stabilised after the Hercynian orogenesis, contrast with these in being more flexible. As we have seen, they have been fairly readily reincorporated in post-Palaeozoic geosynclines. They have also suffered important vertical movements. The Hercynian zone of western and central Europe is typical in this respect. It is broken up into an alternation of ancient massifs where the shield outcrops and sedimentary basins where marked subsidence has permitted an accumulation of 2 000 to 3 000 m of sediment mostly Mesozoic in age. The ancient massifs are distinguished from the shields by: (*a*) their lesser age: they are the product of orogenies at the end of the Palaeozoic, so that they are less indurated; (*b*) their lesser stability, a consequence of their relative youth: they are characterised by larger and more frequent tectonic pulsations; and (*c*) their smaller dimensions, which result from fragmentation due to greater thrusting, a consequence of their lower rigidity.

Finally, there are transitional types between platforms and fold chains. The Appalachians are one such type, close to the platforms in character. This is a region of flexible cover folds emplaced at the end of the Palaeozoic without granitisation, and which subsequently became upwarped into a broad whaleback form at the end of the Cretaceous and in the Tertiary. The Urals present similar characteristics but they were affected by compression again in the Cretaceous because the Jurassic of their eastern margins is vigorously folded while the Tertiary cover is tabular. To what degree have the Palaeozoic folds been reactivated? It is difficult to answer this but, in all cases, differential erosion which has produced Appalachian relief has been influenced by this development, notably by the interplay of Tertiary fault blocks. Appalachian morphogenesis is acting on a superimposed fold structure which lasted well on into the Mesozoic.

The types of tectonic disturbance

These are related, for one thing, to the duration of development, the most ancient platforms being the most stable. The shields, despite their great age, are not generally subjected to deformations of great amplitude. For example, near the margins of the Laurentian Shield remnants of the Palaeozoic cover, notably of Ordovician sandstones, are preserved in grabens. These could have persisted for 350 million years only if the pulses of uplift had been vigorous. In the same way, the Ukrainian Shield is still largely overlain by Cretaceous and Tertiary materials. Like the ancient massifs, therefore, the shields suffer periods when uplift is interrupted and these allow the emplacement of discordant covers in episodic fashion. But, on the whole, they are uplifted very slowly as an assemblage (with the well known exception of glacio-isostatic movements) and are affected by movements of particularly large radius.

The ancient massifs have quite another style. Being smaller, the arching is of a much smaller radius and more often accompanied by fracture and block movement. A number of the less ancient, less indurated and more fragile platforms are fault blocks, horsts or demi-horsts like the Vosges. Thus, in this group, the tectonic uplift appears more rapid, unleashing more intense denudation of the structures and eliminating more quickly the remains of the superficial cover. While these are much more recent than the shields they are no better preserved, often the contrary.

This structural situation, in combination with different types of disturbance, produces several additional varieties of platform. On the margins of the continental blocks, many platforms are warped in the form of a rim. They decline in gradual fashion towards the interior of the continent but, on the oceanic side they plunge sharply by means of faults or flexures which connect them to narrow coastal plains beneath which soundings have often revealed great thicknesses of sedimentary accumulations. These asymmetrical and elongated antéclises are whalebacks (*dorsales*) like the Serra do Mar in Brazil, the Guinean arch in West Africa and the Mayumbe in Equatorial Africa. The remains of ancient fold chains are often encountered, e.g. the Scandes which, in essence, form an uplifted welt about the Fennoscandian Shield. Eastern Labrador constitutes such another arch of uplift (*dorsale*) in the same way. But all welts are not of this unilateral type. Others in the interior of the continents form axes of uplift set between *cuvettes* or broad basins, like the Oubangui dorsale between the cuvettes of Chad, the upper Nile and the Congo. This type is generally less marked.

Arches of uplift and cuvettes, ancient massifs and sedimentary basins occur together in a variety of relationships. They correspond to more ancient and more recent platforms respectively. But poorly differentiated platform also exist where neither the antéclises nor the synéclises are very well marked. In such cases the tectonic rhythm has been spasmodic and discontinuous throughout their development, so that the sedimentary covers are cut by discordancies, and each bed or series has a different extent, transgressing on to what previously had functioned as an antéclise and, in contrast, terminating on the edge of a region which had previously functioned as a synéclise. The Sudan of East Africa or the Russian Plain offer examples of these tables, which are imperfectly differentiated platforms. Tables usually correspond to regions stabilised a fairly long time ago, their evolution into platforms dating from the middle or the beginning of the Palaeozoic. They may be associated with shields of otherwise poor expression, like the Ukrainian Shield, or with arches of uplift as in West Africa.

Types of relief on the platforms

From the geomorphological point of view it is obvious that the difference between antéclises and synéclises is fundamental. In effect, due to slow uplift, downwasting and dissection predominate in the regions of antéclise. The active synéclises are occupied either by epicontinental seas or by large interior plains of alluviation, like the Niger Shield in Mali or the Chad Cuvette. Their fashioning is one of accumulation controlled by morpho-climatic conditions due to the slowness of the deformations and their style. It is not influenced by tectodynamics in the same sense as in the intra-montane basins. The subsidence of the synéclises studied here ended rather a long time ago and they have been relatively uplifted, like the Paris Basin, so allowing the development of a relief of dissection. But, due to lithological conditions, this is very different from the sculpturing of the antéclises. Accordingly, the distinction antéclise–synéclise remains valid and may serve as a framework for our explanation.

Relief forms of antéclises

The relief of antéclises is controlled by:

1. the predominance of resistant rocks which give little scope for differential erosion except in the special case of Appalachian relief;
2. the pulsatory nature of tectonic movements of large radius of curvature which, given favourable climatic conditions, raise the possibility of the extension of erosion surfaces;
3. the rigidity of the shields which results, during major earth movements, in deformation by fracture producing fault blocks.

This distinctive relief in the antéclises has only been fully recognised following recent work; the role of faults, for example, having been under-estimated and that of warping denied at the expense of the eustatic origin of surfaces of planation.

THE ROLE OF LITHOLOGY

The slowness of oscillatory movements on platforms favours gradual dissection. This results in the production of differential dissection forms every time lithology lends itself and climate allows regular evacuation of debris over long distances. In general, these conditions are realised much more fully in uplifted synéclises with a variable sedimentary infilling than on the antéclises which all too often contain rocks indurated by ancient fold chain tectonics and metamorphism. At the same time, among the antéclises there are some which are made up of folded cover rocks that have

169

suffered little or no metamorphism: these favour differential erosion and an Appalachian relief develops on them.

Appalachian relief

Appalachian relief takes its name from a part of the Alleghenies in the eastern United States. A long belt of regularly folded cover rocks constitutes a transition between the granitic intrusions and metamorphic rocks of the Blue Ridge to the east, and the more or less tabular structures of Palaeozoic age to the west, which plunge down into the great synéclise of the central United States. The Allegheny–Blue Ridge ensemble has been raised into an arch of uplift since the end of the Cretaceous and, since that time, has been subjected to an intense dissection under subtropical climates alternately drier and more humid. Successive detrital covers made up of pebbles, sands, and red clays were spread eastwards on the Coastal Plain from the upper Cretaceous to the Quaternary. Several successive major earth movements occurred during this long evolution, the duration of which was about 70 million years. A very advanced degree of sculpture has

FIG. 3.4. Differential dissection of a deep-seated fold (*pli de fond*) in the Cordillera de la Costa, near Caracas in Venezuela
There is a steep drop from the hill-tops to the Caribbean Sea (on the right). An erosion surface may be distinguished, curved tangentially to these ridges. The moderate up-warping has provoked intense stream dissection; the resulting valleys, like those in the middle distance, take on a variety of forms according to the structure; the Cretaceous slates, more or less pelitic and metamorphosed, are vigorously notched (centre and left of photo), whilst a bed of leptynites, quartzose and more resistant, forms a slab, dipping at 70°, on the right; and the main valley is deepened and narrowed

been possible. In the Blue Ridge of Virginia a subtropical type of landscape has been produced with valley heads terminating sharply along the water parting in long steep slopes barely retouched in the cold periods of the Quaternary and which have served, above all, to augment stream action. The Appalachian zone is dissected principally by longitudinal water courses which at long intervals join together to cut across the Blue Ridge, like the rivers Shenandoah and Potomac.

All the essentials of the landscape have been produced under subtropical conditions, periglacial action being scattered and limited to altitudes above 700 m in Virginia. This explains two peculiarities of the dissected relief of the Appalachian zone. First, the behaviour of limestones: these have acted as weak rocks under climatic conditions favouring a predominance of chemical weathering. On those parts of the limestones where honeycombed outcrops frequently emerge above veneers of *terra rossa*, flat bottomed depressions are produced with only a slight recent incision of the streams in the vicinity of the water gaps. Second, the appearance of the water courses, whose longitudinal profile retains very important traces of this dominantly chemical morphogenesis which is little given to mechanical fragmentation of rocks. They are characterised by rapids where the streams cross bands of rock resistant to alteration, notably tough sandstones, quartzites and massive gneisses. These rapids form thresholds which have long prevented incision and have furnished local base levels sufficiently fixed to allow the lowering of the limestones, a process effected essentially by chemical means.

These peculiarities deserve further explanation. In fact, they have been defined as a result of our own observations and do not seem to have been understood by American geomorphologists who have viewed the region while under the spell of Davisian ideas.

In the Appalachians, the whole of the relief is dependent upon lithology. It is characterised by:

1. Elongate crests of massive form with wall-like flanks which dominate the horizon. Their summits are generally flattened and constitute a plateau remnant. These are veritable barrier ridges, the *Appalachian ridges*. They coincide with rocks which are resistant under the subtropical morpho-climatic system, sandstones and quartzites which are not susceptible to chemical decomposition. Being broad and regular, they may be followed for tens of kilometres. The dips are usually between 40 and 60 degrees, with the arcuate ridges of plunging anticlines appearing where dips drop to 15 to 20 degrees. The ridges may coincide variously with the flanks of folds, the bottoms of synclines or the crests of anticlines, although they are made up entirely of siliceous rocks in the latter two cases.

2. Linear depressions, generally broader than the ridges, and called the *Appalachian valleys*. The rivers follow these. Generally, the floors of the valleys constitute local planation surfaces cutting across the structure.

These surfaces are of karstic origin and are strewn with terra rossa, a rubefied clayey sandy formation, being a mixture of residual clays of limestone weathering and siliceous detrital material produced by the neighbouring ridges. The width of these valleys often attains 5 to 10 kilometres.
3. A drainage pattern of trellis type, with long strike reaches within the valleys and short transverse reaches cutting diametrically across the ridges in gorges, the *Appalachian gaps*, usually the site of rapids. As we have already shown, this pattern has developed over a very long time, and explains why such an advanced degree of adaptation of drainage to structure has been possible, which remains evident where drainage changes have occurred. The drainage concentrates at certain points along the resistant ridge in order to cut through it and this has resulted in the abandonment of some of the gaps ('wind gaps' of the American authors as opposed to 'water gaps' which are still occupied by streams).

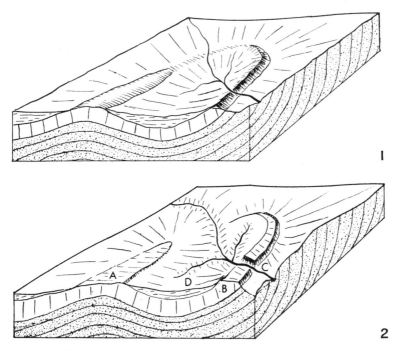

Fig. 3.5. Gradual evolution of Appalachian relief
  1. Gentle folds have been truncated by an erosion surface, and the strata are now subject to differential erosion resulting from a slight incision of the streams.
  2. The evolution continues and the Appalachian features become more marked. The diagram shows: A. erosion of the crest of an anticline; B. an Appalachian cuesta on the edge of a syncline; C. a water-gap through this cuesta, and D. a synclinal valley corresponding to a softer stratum

Successive periods of planation have taken place. The most ancient, which have been uplifted most severely since the middle Cretaceous, truncate the quartzite ridges. There has been a tendency to see in the accordance of summit levels a single surface of erosion, a primary surface, developed before the uplift and deformed in a single event prior to the differential dissection, in conformity with Davisian concepts. This view seems erroneous and some recent studies, unhappily far too few, tend to suggest serious inequalities in the uplift, with warping and differential movement of wedges. Nevertheless, Appalachian relief shows an adjustment of dissection to lithology which is extremely advanced. It is the type region for this process. It demands fairly broad scale tectonics and a long, gradual evolution maintained by a persistent uplift. Thanks to this, the drainage network, by successive adjustment, has been able to adapt itself to the lithology to a remarkable degree. The subtropical climate, in hindering the incision of coherent, chemically resistant rocks has also favoured it.

Such a typical relief can be developed only when a certain number of conditions are realised together: as regards lithological conditions, the regular alternations of bands of rocks with very different responses to sub-aerial processes; and as regards morphogenetic conditions, progressive dissection maintained by the relatively regular and prolonged assistance of up-doming.

It is often the case that one or other of these fundamental conditions is not sufficiently realised. In such cases *embryonic Appalachian relief* is produced, like that of Le Condroz in the Belgian Ardennes. Here, the lithology is suitable, but the uplift has been too recent, namely Plio-Quaternary. It started only 12 million years ago and so has lasted five times less than in the Appalachians. Despite the proximity of important rivers such as the Meuse, the development of the valleys is literally only just beginning and then only around the major rivers. In contrast, gorges mark the passage of these rivers across resistant bands of rock. In the east of the Montagne Noire, events every bit as recent have resulted in even more tenuous relationships in the absence of a sufficiently differentiated lithology: here the resistant rocks are unduly predominant. Furthermore, the folds are much more complex (see the geological map, *Bédarieux*, scale 1:80 000). A fine Appalachian *structure* clearly visible on the geological map, does not necessarily give rise to a very clear Appalachian *relief*.

Pseudo-Appalachian relief and other lithological types

Certain Appalachian structures expressed in the relief do not engender typical Appalachian morphology. In defining the latter, it was seen that the Appalachian ridges bear traces of planations antedating their development even though the planations are not the remains of a primary surface and even when they have been reduced in level by modest movements of structural wedges. This is why typical Appalachian relief is met in certain

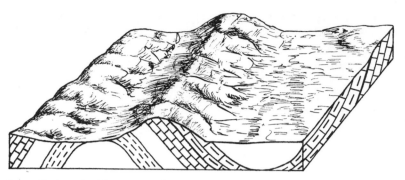

Fɪɢ. 3.6. Appalachian-type relief in central Australia, near Alice Springs
Sandstones, quartzites, greywackes and slates. Under semi-arid conditions,
the softer strata form gentle slopes whilst the more resistant beds stand out
as homoclinal ridges of inselberg form

geosynclinal chains, notably in the regenerated geosynclines: the activa-
tion of a wedge of Appalachian structure allows the sculpture of an Appalachian relief without any difficulty in such cases, as in the High Atlas of
Morocco. Conversely, in many platform regions, these conditions are not
realised and pseudo-Appalachian relief prevails.

The Armorican Massif, and the Taunus and the Hunsruck in the Rhenish
Massif provide good examples of pseudo-Appalachian relief. Thick series
of vigorously folded slates surround narrow synclines of quartzites and
sandstones with a breadth of only a few kilometres. Nowadays, it is considered that, in the first instance, these extremely resistant quartzites
formed limited outcrops only in the depressions and that they have produced
synclines precisely because of their rigidity. Being resistant to deformation,
they would have forced the neighbouring slates to fold around them and to
become uplifted into anticlines. In the event, then, the quartzites only
formed synclines which remained in a perched condition after the stabilisation of the platform. Around Falaise they are seen to give residual ridges of
massive appearance, fossilised by the Jurassic cover and partly revealed
by its excavation (see geological map, *Falaise*, scale 1 : 80 000). In the schists
of the Rhenish Massif only traces of a fossiliferous cover are found whose
origin, moreover, is far from being certain. But the quartzite ridges of the
Taunus and the Hunsruck stand above all the planation surfaces which
truncate the neighbouring slates. Moreover, at Sierck on the German–
French–Luxembourg frontier, similar but very much smaller barrier
ridges of quartzite still make up fossilised relief elements in the Triassic
sandstones.

These pseudo-Appalachian barrier ridges are not simply a stratigraphical
curiosity. Their relief is distinctive and differs from that of true Appalachian
ridges. Having been subjected to prolonged erosion, they are etched away
on their margins, and their boundary does not coincide with the resistant-
unresistant lithological contact, for it has been pushed back into the interior

of the resistant rock outcrop. Never having been truncated by a surface of erosion, they have an irregular summit with massive brows, that is, they have the appearance not of a barrier ridge but of a rounded summit crest (*sierra émoussée*).

Pseudo-Appalachian landscapes appear in the metamorphic series of the shields, particularly under a tropical climate. They are common in Brazil where the *leptynites* (gneisses very rich in quartz) are more resistant to disaggregation than ordinary gneisses and form inselbergs in the form of pseudo-Appalachian crests.

Differential dissection, controlled by chemical weathering, is the general cause of the isolation of sugarloaves, rock domes and lithological inselbergs of equatorial, tropical, subtropical and semi-arid regions. Some of those revealed in the course of the dissection of an erosion surface of regional extent are of Appalachian type origin. Others have always formed part of the residual relief standing above the planation surfaces as far back in time as evidence will take us, and these are of pseudo-Appalachian type. The massiveness of the rocks and, secondarily, the disposition of the joints, play a major role in these relief forms. Post-tectonic phases of granitisation have emplaced massive rocks with rare joints disposed concentrically (*en oignon*). The latter particularly favour the formation of rocky domes under all those climates in which chemical weathering predominates. These domes are met with in all the hot regions, even in the central Sahara (Adrar) where they constitute palaeoclimatic relics, having survived periods of rather more humid climate. They are even found in temperate regions like the Bohemian Massif, where they are the products of tropical palaeoclimates of the Tertiary, and have only been preserved in particularly tough rocks which have survived periglacial attack almost unscathed.

The role of joints being rather more important for chemical alteration than for mechanical fragmentation, it is in the warm or hot-humid climates that the most favourable conditions occur for differential dissection of granitised shields. Those to be seen in deserts such as the Sahara are, generally speaking, palaeoclimatic survivals modified to varying degrees by the disaggregation processes found in arid regions. Under chemical attack in the moister climates, the fissured granitic rocks are readily altered and reduced. In these conditions they are relatively weak rocks, much weaker, for example, than quartzites or even sandstones, particularly the ferruginous sandstones. In contrast, the massive granitic rocks are resistant. Accordingly, relief forms due to differential erosion tend to have rock domes and sugarloaves in massive rocks (granites, syenites) and pseudo-Appalachian ridges in quartzites and sandstones. Nimba, one of the highest summits in West Africa, is a pseudo-Appalachian ridge of ferruginous quartzites. Very often, the tabular sandstone cover rocks make up plateaus standing above the shields in an inversion of relief (Upper Volta; Nigeria; Chapada de Bahia in Brazil). Under the dry climates, granular disaggregation of granitic rocks is intense, with the result that they behave rather as weak rocks in which pediments are readily cut. In southern Morocco, the limestone cover rocks give rise to plateaus which stand above the granitic shield.

Because of the great variety of morphoclimatic systems, differential erosion can act differently in any particular case. The reconstruction of the sequential regimes, which may often have had opposing tendencies as a function of palaeoclimatic oscillations, is therefore necessary in any interpretation of the origin of the relief forms that have been produced. One is reminded of the grave errors which have been perpetrated under the influence of the Davisian concept of 'normal erosion'.

RELATIVE IMPORTANCE OF PLANATION AND TECTONICS

The granitic rocks can be fairly readily reduced to produce an erosion surface under certain morphoclimatic conditions: in this case they yield *glacis* or pediments above which stand the steep sided forms of the inselbergs (literally in German 'island mountains'). This kind of development has been of major importance in the antéclise areas of the shields. Accordingly, we shall study its conditions and then its effects.

Conditions of planation development

These planations are *never* to be seen as Davisian peneplains. They do not form in a humid climate as a result of fluvial erosion. They result from a combination of two mechanisms: granular disaggregation of granitic rocks which produces readily transportable sands or gravels; and intense but diffuse gullying, characterised by the shifting courses of a braided stream system whose concentration is hampered partly by those gravels too heavy to transport and partly by tufts of discontinuous vegetation. This combination of mechanisms operates in a range of climates of fair variety, from tropical savannas where rather more intense decomposition yields more clay and fine products, to semi-arid regions with or without a marked winter. The severe climates of the western United States are particularly favourable due to the importance of granular disaggregation by frost. It is from these regions that we derive the American term *pediment*, applied to these glacis which are uniformly sloping planar surfaces cutting across the structure. They are particularly well developed in the granitic rocks although they are also found in numerous other rocks types when climatic conditions are appropriate.

The glacis develop particularly well on coarsely crystalline granites which are more susceptible to granular disaggregation. Under semi-arid continental climates, like those of the western United States, their development is very rapid. They can even be seen to cut across blocks which are in course of uplift. Pediments often surround subsiding basins of residual geosynclines. The downwarping of these basins facilitates the distant transport of the debris produced so that the pediment has only a thin cover of debris, rarely more than a few decimetres, which is commonly discontinuous. These are the typical pediments. The weakness of the cover facilitates the reduction of any irregularities forming salients. The pediment can even maintain its regularity despite permanent warping, provided that this is not too violent. In other words, it constantly regrades itself. Simultaneously, it progressively attacks the uplands surrounding the basin as long as it is in contact with suitable rock types. Thus fault escarpments retreat without becoming smoothed, forming benches which have often been taken, by those seeking them, for cyclic erosion surfaces.

The formation of typical pediments in granitic rocks is effected by cutting into antéclises which are in the course of slow uplift. This is common in climates particularly favourable to pedimentation as in the western United States. It follows that these planations do not imply a tectonic stability but simply rhythmic uplift which does not exceed a certain rate, so that a dynamic equilibrium with the climate can be established.

In tropical regions, the alternation of humid periods with marked chemical weathering, and drier periods with savanna vegetation or xerophytic bushes conducive to sheet wash and broad-scale removal of the weathering products, favours the development of extensive planation surfaces across the granitic shields. These extensive planations, which are on a much greater scale than typical pediments, are known as *pediplains*. The South African geomorphologist, L. C. King, has emphasised rightly but in rather too narrow a fashion, the essential role that pediplanation plays in the development of tropical shield landscapes. Given a sufficient period of time, this process can effect advanced erosion of the continental arches, and may give rise to level surfaces of regional extent which are disposed about the cuvettes where the detrital formations of sands and clays accumulate.

Palaeoclimatic conditions favourable to pedimentation have prevailed over large parts of the surface of the globe throughout the Tertiary and the Cretaceous, when hot climates with alternating humid and dry periods occurred. Almost every one of the planation surfaces identifiable in the ancient massifs was formed under such conditions. For example, the post-Hercynian surface of the Morvan is a pediplain modified by marine erosion in the Liassic. Beneath the Lias, in fact, strips of typical sandstone are met with. For some 10 to 20 million years in the Permian and the Triassic, a very intense phase of pediplanation seems to have reduced the remnants of the Hercynian chain of northeastern France. This is why numerous planations of antéclises are so regular. In effect, pediplains can develop as readily on granitic rocks as on schists but, in contrast, they have great difficulty on quartzitic sandstones and quartzites, a fact which explains the persistence of inselbergs made up of these rocks. These planation surfaces in no way prove atectonic periods, merely relatively slow deformation which is the most frequent type in platform regions.

Residual relief

Residual relief forms, which dominate the glacis and pediplains, may thus have a variety of origins even though their morphology may vary little. This may happen in relic relief forms corresponding to the advancing edge of the active pediplain as it cuts into the main mass of the antéclise, rather like the migration of a rejuvenation head up the profile of a river, and in fault escarpments in rather uniform rocks which have been attacked during the glacis formation process, and which have differentially retreated.

In the course of their development these two types of slope take on increasingly similar form. They become progressively dissected, lithological factors rendering their retreat less and less regular. After this has gone on for some time, the fault is levelled by the glacis a good way in front of the actual escarpment so that it is difficult to distinguish from an erosional slope. Systematic petrographic studies then become necessary so that lithological differences not at first apparent may be discovered. Because of the relative homogeneity of the rocks, faults within the shields can be levelled; they never lead to obsequent fault scarps. The studies of Beaujeu-Garnier and Bomer in Limousin have reinforced the old explanation of Baulig, Demangeon and Perpillou. They envisage in this region an up-swelling both of the shield and the post-Hercynian surface which truncates it, followed by the development of a series of cyclic erosion surfaces of eustatic origin. In fact, the sedimentary cover never extended much beyond the present limits, and the growth of the antéclise, which has been going on more or less continuously since the beginning of the Mesozoic, has been jerky with the production, by pediplanation, of more extensive planation surfaces. These were deformed and broken in their turn during the Tertiary earth movements. The most elevated massifs are tectonically uplifted blocks, often tilted, on which pediplanation has produced diverse results. The plateaus are the dissected remains of a polychronic surface developed for the first time in the Eogene and then modified in the Miocene after a period of earth movement. The Eogene and Miocene deposits are found juxtaposed at the same level in the Bellac and Confolens regions.

TYPES OF ANTÉCLISE

It is difficult to set up a systematic geomorphological classification of antéclises because of the major role played by palaeoclimatic agencies in their morphogenesis. Relic palaeoclimatic elements are preserved particularly well in the granitic rocks as they are resistant to river incision under warm temperate climates. The massive rock bands give rise to nick-points which persist and considerably inhibit retrogressive erosion. An enormous river like the Congo has had great difficulty in cutting back through the arch separating Stanley Pool and the Atlantic, although its course dates at least from the Miocene.

However, it is possible to say something about some fairly distinctive antéclises.

Sedimentary antéclises

These may belong as much to the great upwarps, like the Appalachians, as to the ancient massifs such as the Ardennes. Formed of folded sedimentary rocks, they have escaped granitisation only because their tectonic history has lacked large scale vertical movements. Also they are always associated with relatively recent fold chain structures, most often dating from the end of the Palaeozoic.

Two varieties may be considered from a structural point of view:

1. Those made up of folded cover rocks, with sufficient lithological variety to provide wide contrasts in resistance over short distances, e.g. the Appalachians. An Appalachian relief tends to develop in this case particularly if quartzitic sandstone strata are present, as these are resistant under almost every kind of climate. The other rocks, whose relative resistance is much more variable according to climate, do not yield typical Appalachian landscapes so readily because climatic oscillations modify the efficacy of differential dissection. The production of Appalachian forms demands a period of time of several tens of millions of years, perhaps between thirty and forty. Only the antéclises with sufficiently prolonged tectonic uplift offer the requisite conditions. Antéclises which have been sharply uplifted or uplifted too recently, such as the Ardennes, yield only an embryonic Appalachian landscape.

2. Those which revive geosynclincal structures are much less favourable to differential erosion, for two reasons. First and foremost, the lithological contrasts are not well developed, the series made up of flysch and essentially fine argillaceous beds being too monotonously uniform. In addition, the greater severity of the ancient deformations facilitated granitisation so that masses of metamorphic rocks interrupt the sedimentary formations as in the Armorican Massif. One can no longer speak, then, of truly Appalachian relief but only of relief arising from differential erosion with a style of dissection on the shales and slates distinct from that on the granitic rocks. Localised bands of very hard rocks such as the quartzites of Falaise or the Taunus can sometimes yield pseudo-Appalachian relief which is generally isolated or circumscribed. The schist or slate formations are readily truncated by erosion surfaces but, by the same token, they are easily dissected into a ridged landscape which is rapidly lowered because the stream courses incise themselves rapidly. That is why it is difficult for the quartzite bands to be trenched by the same planation that levels the slates: the rate of reduction of the landscape is too unequal.

While the antéclises in folded cover rocks are always associated with belts stabilised fairly recently, e.g. the Jurassic of the Ural Mountains, the antéclises made up of geosynclinal sedimentary rocks are occasionally made up of structures stabilised much further back in time, say during the Palaeozoic as, for example, in the Scandes. It is not unusual for them to partially surround a shield area. They must therefore be distinguished as carefully from the morphological as from the structural point of view.

Metamorphic antéclises

These are distinguished from the sedimentary antéclises by a different balance in the composition of the various relief categories.

Landforms due to differential dissection are generally very subordinate. For the most part they are present according to the style of dissection, the density of drainage channels or the profile and average gradient of the slopes, as long as the forms are not hidden by more important factors such as marked differences in morphogenetic processes. Of course, these detailed forms are very susceptible to the effects of climatic oscillations. They become modified very quickly under the influence of changing climates. More permanent are the lithological influences resulting from the type of granitisation (such as the revealing of hornfelsed aureoles due to contact metamorphism, more resistant aplitic veins, occasional veins of relatively massive quartz, poorly jointed intrusive masses) or the excavation of shatter zones which give rise to angular traces along river courses, or rectilinear valley sections whose slopes tend to be very uniform or wall-like (fault valleys).

Planation surfaces are much more important than in the sedimentary synéclises: the granitic rocks favour the development of pediplains and resist valley incision fairly well so that the ancient planation surfaces are not rapidly destroyed. Under dry tropical climates, these planation surfaces are very persistent. In humid climates, or given severe uplift, they develop into dissected hill country (*croupes*). The proportion of the country occupied by residual hills in relation to recognisable remnants of planation surfaces varies as a function of the relative duration of the various successive climatic types and the intensity of tectonic oscillations. In a general way, remnants of planation surfaces are better represented on the margins of the uniformly and gently upwarped antéclises. On the other hand, when the antéclises have been subjected to tilting and fracturing in violent earth movements of not too great an age, such remnants are to be found in the central areas near to the water divide. In effect, in the slowly and weakly uplifted antéclises, the pediplains have time to cut back while the neighbouring regions subside. This is the case in the Ivory Coast with the pediplain which coincides with the so-called 'Continental Terminal' formation. This surface has cut back considerably into the Guinean arch which here is not greatly uplifted. Moreover, slightly warped ancient erosion surfaces can be exhumed over vast distances. This is true of the post-Hercynian surface on the edge of the Ardennes, or on the north side of the Morvan, plunging uniformly beneath the cover rocks of the Paris Basin. By way of contrast, in the sharply tilted Vosges, the remains of the old landscapes of low relief of the Hautes Chaumes (which, it should be noted, represent not a pediplain but a peneplain with residual hills and broad valleys developed in the Neogene) are preserved only on the watershed, thanks to the temporary respite provided by the slowing down of retrogressive erosion.

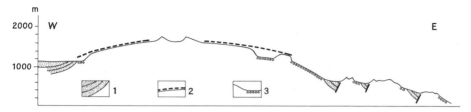

FIG. 3.7. Domed up-warping of an ancient massif: Peñarroya, in Spain (*after Birot, 1959, p. 125*)
1. Detrital material derived during production of summit level
2. Summit level, with several pseudo-Appalachian ridges, deformed by the warping
3. Conglomerates deposited after an early phase of upwarping, and forming a level enclosed within the summit surface. They have themselves subsequently been deformed

In every case residual relief occurs predominantly in the vicinity of the watersheds of the major rivers in areas commonly corresponding with the regions of maximum uplift, in both tilted blocks, such as the Vosges, and antéclises which have suffered simple warping. Slow retrogressive erosion greatly increases their chance of survival, and permits the persistence of old fault landscapes as in the Limousin. It also has the effect of inhibiting development of extensive surfaces of planation: those which form on the edges of the antéclise rarely have time to become well developed in the central portions unless slow subsidence operates as seems to have been the case at the beginning of the Mesozoic in many of our ancient massifs. A well developed pre-Triassic surface has truncated the Vosges, and a pre-Liassic surface has, according to Beaujeu-Garnier, completely reduced the Morvan. However, the considerable period of subaerial attack which has affected these areas, and has lasted until the present, has prevented their preservation: while slow, the incision of the rivers has resulted in their degradation. The watershed regions of ancient rivers are therefore regions of residual relief so placed as to prevent the complete development of planations: they are subject to a slow degradation which produces gentle, rather random hill country where successive modifications due to morphoclimatic changes are attenuated. In these conditions, it is very difficult to discern which areas are those of simple residual relief due to the dissection of a much modified ancient fault block, or even to be sure of the part played by the lithological factor.

Regional tectonics play an important part in the disposition of many types of landform and in their relative importance. Slow, prolonged deformation without severe earth movements favours a gradual development with progressive reduction of relief by planation beginning on the periphery. This has occurred in the Limousin since the Miocene and on the Guinean arch in the Ivory Coast for a very long time. In contrast, any important modification in the action of tectonic components such as has been the case in the Ardennes since the Miocene, is a source of complications. For example, the superimposition of the Meuse provided a lower base level in the interior of the massif which, in turn, accelerated the degradation of the earlier planation surfaces especially on the slaty rocks.

These tectonics are related to the major types of structures. Those parts of the platforms which were stabilised farthest back in time, the shields, are usually (except for recent accidents such as glacio-isostasy) areas of very prolonged, constant but very slow positive tectonic movement. This favours the widespread stripping of exhumed erosion surfaces on their margins, as on the Cambrian surface of Karelia, and the fashioning of undulating landscapes of very slow denudation. For example in Fennoscandia, despite its glaciations, important remains of Tertiary alteration products are still to be found in some areas, testifying to the great age and the little modified nature of the major relief elements. Even leaving to one side the pronouncements of King, it is incontestable that the continental arches of Africa still preserve important remains of weakly degraded ancient landscapes. However, from the very fact of their great age, the

FIG. 3.8. Southern flank of the Aigoual, in the Cévennes
    A granite massif, strongly fractured, attacked under climatic conditions that favour the intense erosion of the fissures by stream action. The various erosion surfaces marked by the ridge-crests have been severely dissected

rigid shields bear traces of the innumerable tectonic events they have suffered. They have not been able to act as unflinching horsts. The fracture lines are innumerable. Some of these have been reconsolidated by minor intrusions which are often harder than the surrounding rock, so that they have been left standing above the surface as dykes. Mineralisation, including the production of precious stones, is generally associated with them. Some other fractures, on the contrary, are shatter zones which facilitate erosion (glacial erosion, fault valleys). The highest incidence of such relationships is met in those fragments of shields which have been incorporated into later orogenic zones, usually of Hercynian age, as in the Bohemian Massif. In the Böhmerwald (Austria), the drainage network displays a suite of rectangular traces controlled by shatter zones.

Relief forms of synéclises

The relief forms of synéclises are very different from those of antéclises. Active synéclises are the sites of plains of accumulation, like the Landes of Gascony and the plain of the Low Countries. To some degree, they escape the implications of structural geomorphology. The relief forms of synéclises that we shall be studying are essentially landforms of dissection which appeared following a reversal of tectonic tendency of variable age or vigour, as we have seen to be periodically the case, for example, in the Mesozoic evolution of the Paris Basin. In contrast to the antéclises, this landscape of dissection is cut into fairly young, essentially tabular sediments. Accordingly, the contrast is very marked.

Besides the processes peculiar to this morphogenesis, we shall examine the specific type landform of the synéclise—the cuesta—and, finally, the forms developed marginal to the shields.

SPECIFIC MORPHOGENETIC STYLES

These result from a variable structural history, with alternations of periods of tectonic depression accompanied by sedimentation, and positive uplift which unleashes dissection. The morphogenetic factors therefore are three: conditions of sedimentation, tectonic evolution and the kinds of recent uplift.

The influence of sedimentation

The manner of the sedimentation plays a very great geomorphological role because it influences the lithology. In the case of subsiding regions occupied by epicontinental seas, sedimentation may be at its most regular. In this case, the facies are controlled by climatic oscillations which usually affect the major surfaces, the very minor relative relief of the platforms essentially eliminating orographic climates. At a subordinate level, palaeogeographical factors may play their part. The disposition of lands and seas controls the emplacement of deltas and estuaries, sandy beaches and dune barriers associated with them, and coral reefs: these can be understood by important facies differences which have their effect on the relief form. Nevertheless it is in the case of sediments of marine origin that the most favourable conditions for the development of a relief of differential dissection are met with, as in the cuestas. In effect, relatively easily eroded beds alternate in the vertical plane and each one has a wide lateral extent. An excellent example of this is provided by the Mesozoic sediments of the basins of London, Paris and Swabia-Franconia.

The terrestrial basins of sedimentation are less favourable to this kind of development because the facies deposited in them are much more variable. The principal axes of accumulation contain coarse deposits while the finer facies are deposited in the distal or marginal depressions where lakes and paludal areas allow slow settling out. However, major climatic changes can stimulate facies variations of regional character. For example, in a humid climate palustral accumulation is of fine clay silts, while in a dry climate accumulation gives rise to broad spreads of gravel and sand beds. In the sandstone shale or clay sand series, major alternations of facies are thus produced and provide the requisite lithological conditions for the development of cuesta landscapes. In tabular regions, where the subsidence is discontinuous, a bed of marine clays may often be covered up again by sands and gravels of terrestrial origin which in time become consolidated as sandstones and conglomerates. This is a frequent situation in coastal plain areas and appears to apply to the Vosges Sandstone. When the beds are sufficiently uniform and thick some very fine cuestas can be developed as, for example, with the 'Falaise' of Banfora in Upper Volta, or the Cretaceous cuestas around Enugu in Nigeria.

However, due to the very conditions of its emplacement, the limestone marl series is in general more likely to develop extensive and regular cuestas than the sandstone shale series with its more variable facies. The limestone marl series is laid down either in epicontinental seas or in large, shallow lakes occupying former marine basins which have become cut off from the sea. This occurred on several occasions in the Tertiary of the Paris Basin (the lakes of Brie, then of Beauce). Occasionally, important sand beds are deposited in the epicontinental seas, as was the case with the Fontainebleau Sands, but this is fairly rare and the sandstone gravel series is usually characteristic of the coastal plains or the accumulations of endoreic drainage basins. The latter are not necessarily caused by subsidence; they can result simply from a markedly dry climate. That is why episodic sedimentation of the flat lands (*tables*) is often made up of this type of material. The Quaternary deposits of the interior part of the Niger delta provide an excellent example.

The influence of tectonic history

The character of the subsidence influences the thickness of beds during the period in which sediments are accumulating. Even without emergence, a slackening of the rate of subsidence can produce *lacunae*, that is to say interruptions or breaks in sedimentation. Dreyfus has drawn attention to these in the Jurassic beds of the Paris Basin and the Jura. In a shallow sea, tidal currents can sweep clear the shallows and so prevent deposition of particles. Sediments deposited earlier can even become superficially hardened and develop a ferruginous coating ('hard grounds'). The progressive reduction in thickness from the Callovian of Lorraine towards the Plateau de Langres is the result of these processes. When sufficiently well developed, they may influence the relief of cuestas.

Generally speaking, in the epicontinental seas the downwarped zones tend to receive a more abundant sedimentation, the finer fractions being more readily deposited beneath the surface zone of agitation where they are kept moving. In contrast, the shallows in clear waters favour, when the climate is warm enough, the development of coral reefs. Where these coral reefs are associated with an advanced recrystallisation of the rock, they are, in very many cases, particularly resistant and upstanding when attacked by erosion. In the valleys of the Meuse or the Yonne, corallian reefs of Sequanian age give rise to sharp, upstanding rock outcrops which contrast with the adjacent surrounding slopes covered with chalky debris. Halts in subsidence give rise to temporary periods of emergence which are expressed as discordancies, like those of the middle Cretaceous on the Jurassic, or of the Tertiary on the upper Cretaceous in the Paris Basin. Often the transgressive beds remain fairly markedly clastic, because they rework the superficial formations laid down during emergence, the residual relief providing the debris. Thus the Albian of the Paris Basin is made up of sands and clays. The Stampian, which is widely transgressive to the northwest of Paris, is composed of sands. Here, the discordant surfaces can be exhumed easily where they truncate harder rocks, in this case limestones. The infra-Cretaceous surface runs from the Ardennes to the Morvan, and plays a major role in the morphogenesis of the Paris Basin. It bevelled the alternating limestones and marls of the Jurassic and so provided the starting point for the formation of cuesta landforms in the Mézières, Verdun and Bar-le-Duc regions.

These erosion surfaces are usually very well developed and attain a high degree of perfection. In fact, the sedimentary basins are generally fairly resistant to uplift so that residual relief is of little importance. Furthermore, they were formed by freshly deposited sedimentary beds, little consolidated and of weak resistance so that they were readily denuded. Often, they persisted close to depressed base levels, those of seas or lakes, which facilitated such attack. Finally, these surfaces are polygenetic. They have been fashioned by several distinct mechanisms in combination: chemical alteration, sweeping by sheetwash, and retouching by marine abrasion in the course of transgression, all these as a function of changing climates and paleogeographies. The infra-Cretaceous surface in the east of the Paris Basin and the infra-Stampian surface in Normandy are both remarkably uniform. For this reason, they bevel the undulating or warped strata. Their dissection provides an excellent starting point for the formation of cuesta landscapes.

The conditions which unleash dissection

Those tectonic phenomena which cause the cessation of accumulation and stimulate land emergence both play an equally important role. They are the origin of the morphogenetic evolution which we can see going on at present. Several different type examples may be cited.

The most simple case is that to be seen along the margins of the Gulf of Mexico in the United States. Slight movement of the continental flexure has gently warped beds deposited as recently as the Plio-Quaternary. In the regions of land emergence, these have been uplifted and very gently tilted towards the sea beneath which they plunge. Thus successive belts of rock are produced which are progressively younger with proximity to the coast, with the exception of those Quaternary formations whose emplacement has been influenced by glacio-eustatism. A surface of regression, dipping towards the sea but inclined much more gently than the strata, is thus developed. It is the starting point for the development of cuesta relief. Also, a drainage network is developed upon it whose principal axes coincide with the slope of both surface and strata. These are ideally simple conditions which are explained by the persistence of subsidence in the

FIG. 3.9. Homoclinal ridges in the alluvial deposits of tilted piedmonts, at the contact of the Andes and the Venezuelan Llanos
Piedmont formations of gravel, in coalescing detrital cones dating from the older Quaternary, tilted at an angle of 20°

Gulf of Mexico and by the fact that, in the Neogene, there has been only marine regression without any other modification of the surface.

Basins commonly occur in which the sedimentary fill is cut across by an erosion surface whose gradient thus differs from the dip of the beds. We have already discussed the example of the eastern part of the Paris Basin which is in this category. Such an evolution yields a superimposed drainage system, giving rise to complications in the course of differential dissection.

Basins with piedmont type accumulations are characterised by a clear tectonic differentiation between the subsiding region and its uplifted border. A good example is provided by the southern part of the Paris Basin. Throughout the Tertiary, a moderate subsidence associated with a clear tendency to uplift in the adjacent Massif Central made of it a piedmont region in which were distributed sandy-clay detrital materials interrupted by episodic deposition of limestones and siliceous lacustrine deposits. According to Gras, the regression surface dating from the end of the Cretaceous has been fossilised beneath later deposits or partly modified and incorporated in more recent planations. Widespread dissection has been relatively recent—Plio-Quaternary: the ancient piedmont is still little dissected and persists on the plateau remnants between the valleys. It has, however, been warped from the Central Massif towards the Loire trench. Certain characteristics are found in this area which are reminiscent of the piedmonts of geosynclinal chains, although in very much attenuated form. In the geosynclines, of course, rapid subsidence allows the accumulation of great thicknesses of molassic sediments which are then more violently warped.

The broadly undulating basins are transitional from cover rock folds. Following light tangential movement, the cover rocks are thrown into upswellings and depressions, as is well shown in the northwest of the Paris Basin. Bray has become the type example of these anticlines of large curvature, which have slight dips and, as evidenced in the *boutonnières*,[1] are surrounded by inward facing cuestas. As is the case in regions of folded cover rocks, these tectonic deformations are gradual and are spaced out over a long period of time. Bray was affected almost throughout the whole of the Tertiary and this has stimulated a progressive dissection of the Cretaceous cover. Being truncated a little more completely each time, the planation surfaces could be perfected. Given a continuation of upwarping, the erosion surface becomes dissected and inversion of relief results: in the first instance, this erosion surface was cut into incoherent formations of middle and lower Cretaceous age. It is a development reminiscent of that associated with anticlinal valleys (*combes*), but the tectonic situation differs in that its radius of curvature is much greater and tectogenesis extends over a much more considerable period of time. However, under the influence of

---

[1] Literally, 'buttonholes': cf. 'slashed-sleeve' valleys in the gently warped Jurassic rocks of the North York Moors. [Editor.]

more violent compression, the folds become more accentuated and the updomings pass gradually into the anticlines of typical cover rock folds. The south of England provides many examples from the Weald, which is fairly gently folded, to the Isle of Wight whose anticlines appear similar to those of the fold belt of the Weser region.

The faulted basins are also the result of a fairly vigorous tectonism. However, these basins are associated with more rigid structures. The Thuringian Basin, embedded in the fringes of the Bohemian Massif, is a good example. This rigidity seems to be related to the fact that the substratum is here a fragment of shield which has been reincorporated in the Hercynian block. However this may be, the basin is elongated southeast to northwest like the neighbouring positive elements, fragments of shield uplifted into horsts. Here the cover rocks are cut up by very closely spaced faults into strips with the same orientation. These tectonics are all the more effective as the infilling of the basin is essentially of Triassic material and therefore of little thickness: this has recorded long-period intermittent deformation since its deposition.

Together with the lithology, tectonism controls the relief of differential dissection and notably the genesis of cuesta landforms.

FIG. 3.10. The 'button-hole' or eroded anticline of Makhtesh Gadol, in the Negev of Israel

In the foreground, pavements of Cretaceous limestone, overlying marls; the semi-arid climate favours differential erosion. The beds dip gently to the right, away from the depression, hence the prominent scarp in the middle distance. In the far distance, the corresponding scarps on the opposite side of the 'button-hole'

THE RELIEF OF CUESTAS

Cuesta (*côte*) relief is characteristic of platform synéclises. Its very definition demands the weakly dipping beds which are so typical of this kind of regional structure.

Characteristics

As in the case of many summits in folded rocks, cuesta landforms are a half-inverted relief. In essence, they develop in tabular, weakly dipping beds under the action of differential dissection which erodes the weak beds on the high points of the folds, the resistant beds at a lower structural level persisting. The surface of resistant rock which crowns the cuesta and so makes up the backslope, overlooks the subsequent valley developed at its foot in the weak subjacent beds which, nevertheless, are situated updip from the scarp. The scarp of the cuesta, cutting across both hard and soft rocks, faces updip towards the anticlinal axis.

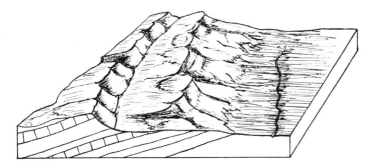

Fig. 3.11. Escarpments in the region of Alice Springs, in central Australia (*after Perry, Mabbutt et al. 1962, p. 69*)
   Limestones, shales and sandstones of upper Palaeozoic age.
A double scarp formed under semi-arid climatic conditions, with 'subsequent' drainage

When ridge forms are eroded in the beds of the cover, lithological conditions are the same as in cuestas. However, the tectonic situation is different. The strata which produce cuestas are tabular. Their dips, as is typical in the synéclises, are of a few degrees only. The cuestas of Lorraine, which are very well developed, have dips of between $\frac{1}{2}°$ and $1°$ only. These slight dips persist over great distances and are the result of tectonic oscillations typical of platforms. Ridges, on the contrary, are associated with folds having stronger dips, of at least 10–15 degrees. Certainly, in undulating platform regions, there is quite a variety of transition from cuestas to ridges. In Bray the dips remain those typical of cuesta relief, except in local tiltings along faults where they are steeper. However, the cuestas match the form of the updoming and are disposed concentrically about an anticlinal valley.

These give way to the ridges of the broad cover rock folds as dips increase, e.g. the hills of the Weser.

The term *homocline* (beds inclined in a single direction) which does not imply a particular dip value, solves this difficulty for us, because it applies to all slopes produced by differential dissection in alternating hard and soft rocks as long as the strata are to some degree inclined. The regional structure is not involved here as the term only applies to local dips.

However, while cuestas pass gradually into ridges, it is important to distinguish one from the other because the dip plays such a critical role in their development.

Structural factors in the development of cuestas

Lithology and structure interact with one another in influencing the differential dissection which gives rise to cuesta landforms.

Lithology acts by itself in two ways: by the relative thickness of the beds and by their variations in resistance. The relative thickness of the beds controls the profile of the cuesta and its relative altitude. The best developed cuestas are those made of a combination of a thick but weak bed and a thin but very resistant bed. This is true of the cuesta of the Moselle between Metz and Thionville. A ratio between these two of four or five to one is particularly favourable as long as the resistance of the hard bed is sufficient. Erosion can then easily reduce the extensive outcrops of the weaker bed, giving a generally concave profile at the foot of the scarp and producing a broad subsequent valley. In contrast, the situation in which the resistant bed is excessively thick allows only a cuesta with a low, rather massive scarp face to develop. It is obvious that a pair of such beds of little thickness cannot give rise to a high scarp. But of course a thick pair of beds provides only a potential for such development: to create a major cuesta it is necessary that the drainage lines be sufficiently incised to allow removal of the weak bed at the scarp foot. The English reader might compare, for example, Lincoln Edge and the Chiltern scarp as specimens of these two types.

Variation in rock resistance controls the mode of differential dissection. In so far as the morphoclimatic conditions allow it, a sharp contrast in resistance gives a concave profile to a cuesta, with a well marked escarpment crest in the resistant bed and long regular slopes with a parabolic curvature in the weak bed. This is the case in regions where sheetwash predominates. In the Mediterranean lands, for example, the rocky crest of the escarpment stands above gradually diminishing slopes which pass into a glacis which merges into the subsequent valley. In contrast, a bed of moderate resistance resting on a not too weak bed gives rise to a broad degenerate slope form especially where solifluction has been dominant. This is true of the chalk cuestas in Champagne. Here a limestone chalk rests on a marly chalk which reduces the lithological contrast to a minimum, although it is just adequate: the cuesta form is attenuated where denudation

FIG. 3.12. Sandstone scarp near Douentza in Mali
   The Ordovician sandstones are of varying hardness. A more massive
outcrop creates an abrupt scarp feature, the beds dipping at about 12°. The
climate is dry tropical, with diffuse surface drainage. The scarp begins to look
like an inselberg, with a steep (30°) and straight slope, above a sandy *glacis*

is weak, for example around the water partings where there are only small
streams. A smooth, broad convex slope passes towards its base into a
gently concave slope cut into its foot.

   The dip influences the mode of dissection and downwasting. Usually,
the resistant beds of the cuestas are also the permeable beds, the weak beds
being impermeable. In subhumid and humid climates, some of the water
infiltrates into the resistant bed and its subterranean passage is controlled
by the dip. When the latter is strong, subterranean flow towards the backslope is important and water escapes rapidly from the cuesta scarp. It also
supplies springs on the flanks of incised valleys developed in the backslope,
when they are sufficiently deep to reach the base of this hard stratum. If
valleys of this depth are lacking, the structure retains the water in an artesian condition. In that case, there are few springs on the scarp face. Its
dissection is then the work of the geomorphological processes of the interfluves: surface wash and soil creep. It tends to retreat slowly and in a fairly
regular fashion. In other words, the attack upon the scarp is fairly uniformly
distributed. The scarp front presents a little-indented, massive appearance.

   In contrast, there is the situation where dips are slight. In this case,
subterranean flow is no easier down the dip than it is towards the scarp
face. A gentle rise in the water-bearing bed favours such flow throughout
the area in close proximity to the scarp face. Thus, some springs can function at the scarp foot or, more usually, a little further down, because the

water continues to move below the surface beneath the gravels and slope waste shed by the scarp. These give rise to rivulets which effect an irregular dissection. Some gulleys run in a direction contrary to the dip of the beds, from the scarp face towards the subsequent depression (*obsequent valleys*), cutting the front of the cuesta into ridges and promontories, most often from 1 km to 3 km in length. Anomalies in the local dips influence the degree of development of these valleys. For example, their development is usually more advanced along faults, the subterranean water flow being impeded by the discontinuity of the beds. Thus valleys which run into a fault tend to extend themselves more rapidly than others, by growth along the fault trace. Hence the cuesta face can be dissected especially if the resistant bed is particularly tough. *Buttes* or scarp outliers, capped by a remnant of the resistant bed, thus become isolated in front of the main scarp. The length of time they persist is directly proportional to the degree of resistance of the caprock.

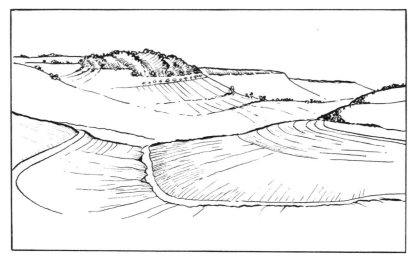

FIG. 3.13. Muschelkalk scarp near Hardheim, Alsace (drawn from a photograph)
   Concave profile, but lacking the usual sharp, cliffed top owing to periglacial action in the Quaternary. A subsequent valley, and moderate scarp dissection by obsequents

   Although, in most cases not very vigorous, synéclise tectonics will evidently play a major part in the development of cuesta relief. As we have shown in relation to the eastern part of the Paris Basin, the geomorphologist cannot afford to neglect this fact. Given that lithological contrasts are strong, fault zones are particularly favourable to the dismemberment of cuestas into groups of buttes. In the Thuringian Basin, particularly advanced tectonic disruption of the structure produces very foreshortened cuestas, frequently interrupted by faults, with an abundance of buttes.

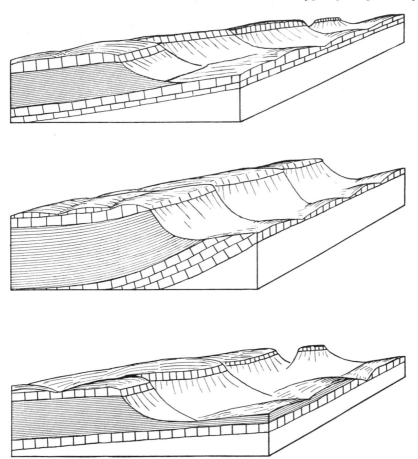

Fig. 3.14. Relief of escarpments

In the top sketch, beds gently inclined, with the scarp carved into promontories by consequent gaps. The subsequent scarp-foot valley is well-developed, and a structural surface appears below the scarp, formed by a limestone bed lying underneath the marls that form the scarp-foot vale.

In the middle sketch, the scarp coincides with a flexure; a straighter feature with no gaps. The subsequent valley is restricted owing to the steeper dip of the strata.

In the lower sketch, the strata are very gently inclined, and the scarp is cut by gaps that in one case have isolated a scarp outlier (*butte témoin*). The subsequent vale is wide and developed entirely on the marls, except where, near the *butte*, a more vigorous stream has cut down to the limestone beneath

In the more uniformly warped synéclises, like the eastern Paris Basin, any variations in dip are geomorphologically very important.

Undulations in the strata, in modifying the direction of the dip, have an effect on the plan form of the cuestas. With partial inversion of relief, the cuestas constitute positive relief elements within the synclines, like the

Grand Couronné on the Côte de Moselle near Nancy or the Toul salient on the Côte de Meuse. Conversely, the negative relief elements, the re-entrants, develop on the anticlines, for example at Pont-à-Mousson and at Commercy. When the structure is one of more severe undulation, the variations in the cuesta's outline are themselves more marked. It is for this reason that they are much clearer in the middle Jurassic around Nancy than in the Cretaceous Champagne *humide*. In fact as we have seen, these undulations are derived from the shield and are transmitted more faithfully the thinner the overlying cover rocks. The outline of cuestas developed in basins with a thin cover or those severely affected by folding, is always more sinuous and more variable. In contrast, the more rigid tabu-lar areas yield very regular cuestas. This is the case in palaeozoic rocks in Mali and in Upper Volta, or in the Ituri region of the Congo.

Any variations in the dips, such as the alternation of steeply dipping limbs and zones of low dip (*paliers*) typical of the simple monoclinal fold, are also important. They are very clear in Lorraine. Having very slight dip, the *paliers* favour dissection which is accentuated by the development of obsequent valleys. In such cases, the depressions of subsequent type are broader because of the lower average dip which allows the weak bed to extend further in front of the scarp foot. The cuestas are thus well developed but somewhat disjointed with isolated buttes, especially if faults are present so that the scarp becomes dissected into promontories and spurs. In contrast, flexures, with their steeper dips, prevent dissection. The extension of obsequent valleys is arrested, while the buttes and spurs which charac-terise the *palier* on the updip side of the flexure are progressively worn down. Slowly, the cuesta becomes aligned along the flexure where, being propped up, dismemberment and retreat cease. Further dissection must await the attack upon the reverse slope: this is effected by headward erosion of valleys whose springs have a high discharge due to the fact that sub-terranean flow is facilitated by the steepness of the dip. These peneconse-quent valleys eventually breach the crest of the cuesta by headward erosion. Once the resistant stratum is breached, the peneconsequent streams are no longer fed from the valley head but from springs situated along their lateral slopes. However, the threshold, now made up only of weak rocks separating the dipslope stream heads from the subsequent valley, may be gradually reduced by slope processes. The resemblance to the transverse gap of a consequent stream becomes more and more marked. But the stream itself has its source a little downstream from the cuesta scarp. It is for this reason that, on many occasions in the past, such peneconsequent gaps have been taken for consequent valleys beheaded by river capture. In such cases, however, there exists no trace of ancient surface flow to prove it: often no consequent valley can be projected further upstream across the subsequent valley. The confluent valleys of peneconsequent gaps, fed by sources on the backslope, retreat little by little towards their heads and so progressively dissect this backslope. (Many examples of such a sequence can be observed on the geological map, *Nancy* and *Commercy*, 1 : 80 000 scale.)

It is important to note that there may be a degree of mutual compensation between structural factors and the morphoclimatic conditions. For example, the effect of dip which we have just analysed is limited to fairly humid regions so that artesian structures exist and the flow of water is not at all seasonal. However, in the dry lands, where surface wash predominates, surface process and structure reinforce one another. In effect, the loosening of debris and its distribution along the foot of the scarp is easier as the dip becomes gentler. Given that erosional attack is similar in both cases, a cuesta with gently dipping beds has a gentler scarp crest than a cuesta with steeply dipping beds. Gulleys undermine the latter more easily and dissect it more because the removal of debris from the scarp crest by concentrated wash is more easily effected.

The resistance of a bed can be to a large extent compensated by its impermeability. Even when very coherent, the impermeable beds are only moderately resistant. Gullying attacks them everywhere and dissects them, particularly when this bed is only moderately coherent. When the bedding is almost horizontal, it is slightly more resistant because surface wash is much less concentrated and its erosive potential reduced. This is demonstrated in the cuesta of the Infra-Liassic Sandstone of Lorraine, which has been reduced to a series of plateau remnants, and the buttes of siliceous limestone in the Paris region, notably those of Oligocene (Aquitanian) age. On sands and subjacent marls concave slopes are produced, although these forms are limited to the plateaus which coincide with the *paliers*. With inadequate dip and such advanced dissection, we gradually leave behind the province of typical cuesta landscape for that of residual plateaus which are fairly common in the weakly disturbed synéclises: barely warped or folded tabular elements, or the central parts of basins overfilled with sediments, e.g. the Paris region where the cuestas are only weakly outlined, even where the lithological contrasts are sharper, due to the dips being too gentle and, at the same time, too variable.

The importance of the different methods of dissection

Davis formulated an explanatory scheme describing the way in which cuesta relief develops. He assumed a simple type of emergent surface, like that of the coast of the Gulf of Mexico. This surface sloped towards the sea broadly parallel to the dips of the underlying rocks. The result was a river network made up initially of trunk streams flowing towards the coast and running in the same direction as the dip. These were termed *consequent streams*. Once the weaker strata had been sufficiently exposed, the subsequent streams were added to the network, and then the obsequents. This theoretical model is a special case and is not the usual one. However, the terminology is so well established, that it may be retained as long as we do not retain the genetic implications, i.e. as long as we do not consider the consequent elements as older than the subsequent or resequent components (for the latter cut through the cuesta scarp many times over). We have already seen in the eastern part of the Paris Basin, for example, that the resequent streams (Moselle, Meuse) are the oldest and that it is only very recently that some river captures have allowed the broadly consequent drainage to increase its catchment at the expense of the Seine basin. This terminology is useful if the criteria employed are not those of an evolution which often contradicts theory: it should express the relationships between the direction of water flow and the dip of the rocks. In this case, consequent streams are those which follow the dip, and obsequents those which flow against the dip, while subsequent streams flow along the strike and resequents cut across it.

In every case where serious palaeogeographical reconstructions have been made, cuesta relief has been shown to result from erosion surfaces cutting across the structure. These include surfaces of retrogression (the plain of the Gulf of Mexico), fossilised and exhumed erosion surfaces (cuestas of the southern part of the Paris Basin, of the Barrois and at the southern foot of the Ardennes) or crude erosion surfaces truncating the strata following a relatively prolonged emergence. It is for this reason that it is quite rare for the backslopes of cuestas to be structural surfaces (true dipslopes). Usually, the backslope cuts across the resistant stratum and is derived, more or less directly, from the initial erosion surface. Around Verdun, it can be shown that the cuesta in the Kimeridgian is the result of exhumation and the dissection of the infra-Cretaceous surface bevelling the upper Jurassic. As differential dissection is still poorly advanced, the surface is well preserved. Often this is not so, and degradation has transformed the surface into a ridge and valley landscape from which the characteristic deposits have been removed. This is the case in Lorraine where the initial Oligo-Miocene surface has to be reconstructed. Limestone plateaus of Hercynian Europe have been strongly affected by Quaternary periglacial action which has markedly degraded them.

The general occurrence of a high degree of conformity between erosion surfaces and the strata they cut across, the existence of discordant sedimentary covers, and slow warping during dissection all tend to multiply the phenomena of superimposition. Several varieties of cuesta landscape result from this.

Twin cuestas (*côtes dédoublées*) appear when, in order to reach the subjacent weaker stratum, the stream becomes fairly deeply incised into the backslope. This produces two parallel asymmetrical slopes composed of the same strata: the scarp face proper and the slope of the developing valley which faces upslope. Its other face, broadly conforming to the dip, develops no scarp, just an ordinary slope. This kind of situation arises in several ways. On a local scale, a small river may be able to take advantage of the fault running parallel to but behind an escarpment and so erode it, as with the Eparges in the Côte Bajocienne not far from Verdun. Such a situation, however, is only an anomaly in the drainage network and only affects a short sector of the scarp. On a regional scale, a resequent river may be found to flow fairly close to the scarp over a considerable distance and, in order to reach the less resistant beds beneath, to become fairly incised. A good example of this is the Moselle between Frouard and Metz. In contrast the Meuse, in failing to reach the Oxfordian marls, is unable to split the Oxfordian cuesta even though it runs along its backslope and parallel to its scarp throughout the distance from Verdun to Commercy.

All the evidence suggests that such dispositions favour rapid dissection and dismemberment of the scarp. In effect, the tributaries which come from the updip side have the advantage of a good water supply from springs or from surface flow according to the particular case. These streams become established in the weaker strata, once there is a certain degree of incision, due to the gentle dips and so remain there for the greater part of their course. The Côte de Moselle (or Côte Bajocienne), between Frouard and Metz demonstrates this very well (see the geological map, *Commercy*, 1:80 000 scale). To the south, in the Toul syncline, the cuesta extends farther to the east swinging across the line of the Meurthe–Moselle drainage. It is cut into strips by aligned peneconsequent rivers which happen to have extended their heads precisely to the line of the scarp. Towards Pont-à-Mousson, by way of contrast, an anticline gives rise to a re-entrant which comes very close to the scarp of the Moselle. Here the dissection is more advanced and only a series of buttes is left between the Moselle and the scarp with broad gaps between them.

Twin cuestas must not be confused with *double* cuestas which are simply two superposed cuestas, a single slope being made up of two pairs of beds. The existence of a double cuesta implies differential scarp retreat. For this to be possible, the cuesta arising from the lower of the two pairs of beds must be overtaken in its retreat by the upper cuesta. Now scarp retreat is only a special case of the retreat of slopes. For a scarp to retreat without attenuation and without disappearing, a certain degree of parallel retreat of slopes is needed which, in turn, requires certain special climatic and lithological conditions which will be examined later. They are by no means always realised. For example, in temperate climates, reduction of cuestas is, above all, by means of discordant dissection, the formation of peneconsequent water gaps, dissection of the backslope into plateau remnants, isolation of buttes, and, finally elimination of the latter once the resistant cap rock has been removed by the retreat of slopes. In these conditions, a valley which is relatively incised into the backslope of a cuesta in a resequent situation can result in the appearance of a double cuesta. This is true of the Meuse to the north of Neufchâteau. In this case a double cuesta and a twin cuesta appear together. But the twin cuesta has produced the double cuesta. In essence, the remains of the twin cuesta on the right bank of the Meuse have been progressively reduced to the condition of isolated buttes.

On the left bank of the river, the scarp picks up again where two pairs of strata come in, namely Liassic marls and Bajocian limestones at the base (which gives rise to the twin cuesta) and Oxfordian marls and Rauracian limestones above, which previously dominated the valley. Standing above the level of a major river the Oxfordian scarp has been very slow in its retreat, while the dissection and removal of the backslope of the Bajocian twin cuesta on the right bank of the river has been easy. The two slopes have therefore become compounded into one with, of course, two-tiered scarp crests.

The tendency is, therefore, for streams to incise themselves *in situ*, whatever the structure. They become extensively superimposed and this is made all the easier if beds are not too resistant to incision. River captures may eventually amend the most marked of the anomalies. But these are produced only after a fairly long evolution, as is shown clearly in the eastern part of the Paris Basin. Essentially, the principal rivers provide a local base level for the smaller streams and in turn to the processes fashioning the slopes. That is why we prefer, in characterising this combination of processes, to speak of differential dissection and not differential erosion, because the latter expression stresses too exclusively the action of rivers. Whatever the initial disposition of the rivers of an area, a cuesta landscape can develop as long as the structure is favourable. But if the entrenchment of the rivers

FIG. 3.15. A composite scarp-face in the Vich basin, Cataluña, Spain
Marls, chalk-marls and limestones, of Mesozoic age. The scarp-face has been rejuvenated by man-induced obsequent ravines that have exposed the strata; the cliff at the scarp-top is very well developed. At the foot of the screes a small subsidiary scarp has formed on a bed of limestone in the chalk-marl series

ceases in a humid climate, slope processes fragment and round off the cuesta forms, the slopes becoming degenerate. Colluvium accumulates at the foot of the slope as the scarp crest becomes rounded. The sharpness of the Lorraine cuestas is due to the maintenance of stream erosion through the Quaternary, itself controlled by uplift. The early Quaternary terraces of the Moselle, the Meuse and the upper Marne stand 60–80 m above the valley bottoms. The development of cuestas during the Quaternary has thus been important. They were much less well developed at the end of the

Fig. 3.16. A scarp produced by limestone resting on marl, in the Agenais region of France
    A well-marked scarp of limestone, little affected by Quaternary gelifraction. The marls form a belt of cultivated land, with a convex slope resulting from soil-flow in front of the scarp

Pliocene, before the effects of the Plio-Quaternary uplift had unleashed this incision. Certain cuestas not yet affected by retrogressive erosion bear the marks of this, e.g. those developed in the Portlandian in the south of the Barrois. Along the water partings, this area is dissected only by small streams which were barely entrenched in the Quaternary. The result is a confusion of poorly aligned rounded hills, often with convex slope profiles, making the cuesta outline sometimes difficult to make out.
    The effects of these same structural and morphoclimatic factors are to be found in the landforms at the contact between sedimentary cover rocks and the shields below.

LANDFORMS DEVELOPED AT THE CONTACT BETWEEN COVER ROCKS
AND SHIELDS

The lithological differences are generally clear at the contact of shields
and their cover rocks. Also, these contacts are particularly favourable to the
development of some distinctive relief forms incorporating some geomorphic types of the shield with those of the sedimentary cover. It is these
diverse geomorphic associations that we shall now study.

The geomorphic assemblages of the shield contact zones are controlled
initially by differential earth movements which give, on the one hand, an
antéclise which is the shield and, on the other, a synéclise, where the
covering rocks are deposited. These contact regions are thus true geodynamic hinge lines. The geomorphological assemblages that are met
here are, therefore, dependent in the first place on the way this hinge line
has acted in the past and, on the other, the nature of the successive morphoclimatic systems which have prevailed.

Contacts beneath discordant deposits

This type of contact is the only one whose characteristics are dominated
by epigenetic processes, and for this reason it is dealt with first. The contact
between shield and cover rocks is masked by surficial debris and by the
glacis. The contact is not expressed in the relief but made up of monotonous
plains where erosional forms such as pediments alternate with forms of
accumulation.

Such an assemblage of forms demands certain climatic conditions. It
can develop only if sheetwash and granular disaggregation are strongly
predominant. Accordingly, it is encountered in essentially dry climates, in
the steppes, savannas and areas covered by xerophytic shrub and bush
vegetation. It is particularly extensive in Africa. Around Banfora, in the
south of Upper Volta, the granitised Precambrian shield of the Guinean
arch shows this type of relief where it comes into contact with the Palaeozoic cover. The contact itself is not precisely expressed topographically,
for a slightly dissected pediplain truncates the cover of sandstone and shale
which ends in a featheredge. It is only some 10–20 km beyond the contact
that the imposing cuesta, with residual relief forms, appears: this is the
'Falaise' of Banfora, composed of Ordovician quartzitic sandstones. On
the south side, the relief of the arch is extremely monotonous, showing only
slight dissection and containing small remnants of cuirassed (duricrusted)
plateaus, the vestiges of more elevated ancient surfaces.

As a result of climatic changes, such contact regions can become dissected.
In the initial stage, only more or less incised valleys are present, there being
little differential removal of the cover rocks. For example this may be seen
around Abidjan in the south of the Ivory Coast, the upwarp being unilateral in this case. The arch descends sharply towards the depths of the Gulf

of Guinea in a continental flexure. A narrow strip of Cenozoic and Mesozoic sedimentary formations thickens rapidly beneath the lagoons and towards the open sea. On the Ghana frontier, several hundreds of metres of upper Cretaceous marls and limestone marls have been encountered in a borehole sunk during petroleum reconnaissance. A submarine canyon, the Trou-Sans-Fond, demonstrates the working of this flexure. At the surface, the older formations, including the shield, are buried beneath the formation termed the *Continental Terminal* made up of deposits of a dry, tropical climatic regime including clayey sands, clays, strings of quartz pebbles, lenses of duricrust and sporadic beds of sandstone. Emplaced during the Tertiary while the flexure was still active beneath the present coastline, it ends on the shield in an irregular featheredge and thickens rapidly towards the coast. The top of the infilling, a surface of accumulation, passes into the glacis in a thin film of deposits on the shield. The great difference between this area and the region about Banfora, is that this surface was developed in an unstable region while deformation was actually in progress. That is why it embraces both the erosional forms on the uplifted shield and accumulation forms in the region of subsidence. This is very common when the margins of the shields coincide with a continental flexure. The southwest of Mauretania belongs to the same type.

In the lower Ivory Coast, after the deposition of the Continental Terminal formation, the continental flexure was displaced towards the open sea and the plain burying the contact was uplifted in a block which, in turn, stimulated its dissection. This dissection is limited, for the present, to valleys incised into the sedimentary fill. Retrogressive erosion abuts on to the shield, where the streams develop rapids. They have not yet reached a stage where they are sufficiently established for a depression to have opened out in the Continental Terminal at its contact with the shield.

A history similar to that of the lower Ivory Coast can be seen in the United States around Baltimore and Washington, where the contact between the

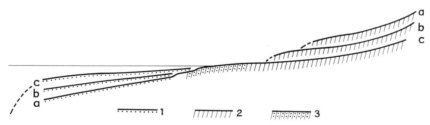

Fig. 3.17. Evolution of a coastal platform under the influence of continental warping (*after Mescheriakov, 1960, p. 32*)
1. Successive deposits on the sea-bed, on the tilting continental platform.
2. Erosion surfaces on the uplifted side of the platform
3. Neutral zone of base-level coastal plains, re-graded during the evolutionary process
The letters a, b and c denote three stages in the evolution

metamorphic shield and the coastal plain disappears beneath a surface partly erosional and partly depositional in nature which is dissected into a plateau. The fall line corresponds to the point where the stream bed leaves the crystalline rocks to attack the Cretaceous and Tertiary sedimentary fill making up the coastal plain.

In the south of the Paris Basin, on the edge of the Massif Central, similar shield margins are to be found, the shield having been truncated by a pediplain and the Mesozoic cover rocks re-covered by ironstone formations emplaced in a typical sheet wash regime. Since the middle Tertiary, weak uplift has stimulated some downcutting of the rivers, although this has been insufficient for the removal of these beds. Where this activity is more advanced, isolated structural relief elements appear in the subjacent Mesozoic cover. (For a good example of the conjunction of shield margins and a glacis see the geological map, *Aigurande*, 1 : 80 000 scale.) Thus we progress, by numerous stages, towards the other types of contact dominated by structural factors.

Shield margins with asymmetrical peripheral depressions

This type has come in for a good deal of attention from geomorphologists, and the Terre-Plaine, on the northern edge of the Morvan, has become the classic example. However, there are clearer examples, such as the edges of the Sierra de Guadarrama near Segovia in Spain, studied by Birot.

A shield margin with asymmetrical peripheral depressions is made up of three topographical elements: an elevated region which gradually becomes separated from the depression and then develops, some distance from the latter, the highest altitudes in the whole region; a depression having the same configuration as a subsequent valley at the foot of a cuesta with a gently undulating, soft relief, then a gradual ascent in the form of an inclined plateau becoming increasingly hilly with proximity to the elevated region noted above; and a cuesta with a backslope whose surface stands above the depression. The summit of this cuesta is normally lower than the elevated region.

From a structural point of view a shield margin with asymmetrical peripheral depressions brings together two geomorphic assemblages: a cuesta relief developed in the covering rocks, giving rise to the depression and its bordering slope; and a system of planation surfaces cutting across the shield. One of these is an exhumed, discordant surface which plunges beneath the cover rocks and which, revealed from beneath the latter, rises gradually from the subsequent depression towards the shield. It makes up the margin of the depression. The other, generally more degraded surface, cuts across shield and cover rocks alike, crossing the summit of the cuesta and passing above its backslopes.

Such a situation is simple and corresponds to the ordinary tectonic character of the platforms. A transgressive cover is laid down on the edge of the shield during the period of diminishing uplift or of momentary down-warping; this is then elevated and warped into an antéclise with the shield during a renewed uplift. In a further period of reduced uplift or of favourable climatic conditions, the warped shield and cover rocks are truncated by a planation surface. With renewed uplift, together with a probable change in climate which serves to increase the concentration of the runoff, this warped surface is attacked by stream dissection which becomes differential in nature.

It is at this point that lithological conditions become of major importance. Well marked contrasts are necessary, bringing together:

1. A resistant shield. The granitic rocks provide the best conditions because they are conducive both to pediplanation under a climatic regime favouring granular disaggregation and sheetwash, and because of their resistance to fluviatile dissection. In contrast slates, too readily dissected and broken down into clays, are not so conducive: the surface of discordance in the cover rocks cannot be exhumed in good condition.
2. A cover made up of pairs of alternating hard and soft rocks such as those which give rise to cuestas. The soft rock must be marly or clayey for differential dissection to be fully effective. Clearly, if the shield is schistose, the contrast will prove inadequate for the surface of discordance between cover and shield to be clearly exhumed.

In the Terre-Plaine in front of the Morvan, these lithological conditions are not entirely realised. Here, the oldest bed of the cover resting on the granitic shield is made up of Liassic marly limestones which are not readily removed. However, they are of little thickness and they are soon masked by a thick series of marls, giving rise to a subsequent depression at the foot of a cuesta crowned by middle Jurassic limestones. The flanks of the depression on the ancient massif side constitute a structural surface and not an exhumed surface of discordance. It may be seen, therefore, that the typical asymmetrical peripheral depressions demand a series of structural and morphogenetic conditions in which palaeoclimates play a part. In the Hercynian zone of Europe, the tropical climates of the Tertiary facilitated the development of planation surfaces which truncate both shields and cover rocks. They have produced the glacis type of shield margins previously mentioned, whose dissection under the more humid climates of the Neogene and then the Quaternary has allowed relief forms of differential erosion to appear. From Aigurande to Avallon, the northern border of the Massif Central provides almost the whole gamut of transitional forms controlled by a fairly advanced degree of dissection of the Eogene glacis.

A variant of the asymmetrical peripheral depression is frequently encountered in tropical regions. Due to the climatic conditions, the intensely weathered granitic rocks behave like rocks of only moderate resistance. In contrast, the sandstones and particularly the quartzitic sandstones and sedimentary quartzites may be highly resistant to attack. Sandstone cover rocks, made up of Eocambrian and Palaeozoic beds, are often found to have been laid down in the interior basins of the shields. In such conditions, warping may produce a differential dissection which ends in an inversion of the relief which becomes progressively more marked as the landscape evolves. The rotted granitic shield is eroded into a depression while the sandstone gives rise to a scarp which is more pronounced the more re-crystallised and massive the beds become. This is a pseudo-cuesta, its resistant bed being the sedimentary cover and its weak bed the shield. In drier periods the latter may be fashioned into a glacis at the foot of the cuesta. The Banfora region, already cited, provides a good example of such inversion of relief, although the cuesta is entirely sedimentary. However, the summit of its scarp is much more elevated, by some 700 m, than the uplifted arch of the shield to the south of it. In Brazil, in the state of Bahia, the Eocambrian tabular beds of sandstone and quartzite make up a sharp cuesta standing above the shield in certain parts of the margins of the Chapada. The whole of the Chapada, a region of cover rocks, is a regional inversion of relief with respect to the crystalline dome. In fact, the Chapada reaches 700–1 000 m in altitude against the 200–400 m of the crystalline province.

Shield margins with discontinuous depressions

A series of types are grouped under this heading where, for a variety of structural reasons, a typical peripheral depression has not been etched out. These, then, are discontinuous depressions loosely separated one from another in a string, or isolated basins.

1. *En echelon* depressions. The southern border of the Ardennes is a good example. Here, a series of cuesta landforms runs obliquely against the edge of the ancient massif. This situation is the result of special conditions during the emplacement of the cover rocks. During the Mesozoic, the Ardennes were affected by tilting movements. In the Triassic and the Jurassic, they were gradually lowered on the west side, while the extreme east was up-lifted again after the Trias. As a result, the Mesozoic seas gradually over-lapped on to the massif in the southwest. Their deposits transgressed obliquely on to the shield. Each limestone stratum ends in a featheredge, reaching a little farther than the subjacent marly bed.

Where a marly bed pinches out against the shield, the latter, although slaty, has been scarred and the surface of discordance is exhumed. A peripheral depression, dominated by a cuesta, is produced. However, the limestone bed of this cuesta, towards the northwest, transgresses directly on to the shield: the peripheral depression then disappears in the absence of a weak bed. The backslope of the cuesta runs directly on to the ancient massif. Accordingly, each peripheral depression dies out tangentially against the margin of the shield at its extreme northwestern end, and then passes, towards the southeast, into a simple subsequent depression as soon as another limestone stratum of greater age interposes itself between it and the shield.

As a result of earth movements, there are often divergences from the classic asymmetrical peripheral depression. In such cases, transgression of the cover rocks is of the direct, frontal type because there is virtually no displacement of the antéclise. In the case of *en echelon* depressions, transgression was oblique due to a migration of the centre of the antéclise related to a seesaw motion of the ancient massif.

2. Depressions caused by the excavation of structural basins. Infilled, tectonically active basins of the final phase in the geosynclinal cycle may sometimes be incorporated into an antéclise once the region has been transformed into a platform. With their sandy clay or sandy gravel sediments, they provide a lithological contrast with the neighbouring shield, especially when the shield is composed of granitic rocks. Under suitable climatic conditions, differential dissection may then proceed. Thus the sediment-filled basin is partially excavated, yielding a depression on the edge of the shield.

Cases of this type are common on the edges of the French Hercynian massifs. For example, the Lodève Basin (southern Massif Central) has a thick fill of Permian mudstone-shale, which has fossilized the residual relief elements of the basement. Uplifted together with the extreme eastern end of Black Mountain, it has been partly removed by differential erosion to produce a depression at the mountain foot. Towards the west around Lunas (see the geological map, *Bédarieux*, 1 : 80 000 scale) it is overlain by a very resistant tabular cover of calcareous sandstones of Triassic age and, to the south, by a Palaeozoic horst on which the Mesozoic is sharply discordant.

The Autun Basin is in the same category, as is that of St Dié. The latter, tectonically more simple, is particularly typical. A localised basin, filled with Permian sandy mudstones, is the site of a late Hercynian depression surrounded by crystalline rocks. The Triassic sandstone cover has transgressed as much on to the basement as on to the Permian infilling. The recent uplift of the Vosges has caused the dismemberment of the Triassic cover. Once they were sufficiently downcut, the rivers worked more readily where they came up against the Permian, thus allowing wasting of the slopes by removal at the base. On the site of the old Permian basin, a good deal more of the sandstone has been removed there than where they directly overlie the resistant basement. Within the basin, they have been reduced to buttes between which the Permian is partly excavated whereas elsewhere they make up massive plateaus.

Volcanic rocks intercalated with the less resistant basin fills introduce an additional complexity of detail into the working of differential erosion. This is the case with the German Permian basins of the Sarre and the Nahe. The wealth of forms becomes greater with each structural complication.

3. Faulted contacts: these are in some ways analogous to the *en echelon*

asymmetrical depressions. They also give rise to strings of successive depressions. When a fault affects both a shield and its covering rocks, the latter are rapidly removed from the shield when it is uplifted. When an erosion surface truncates both and levels the fault, lithological differences are not uniformly distributed. On the upthrow side of the fault, the shield is homogeneous or almost so, and especially resistant when it is granitised. On the downthrow side, the truncated beds vary from one point to another according to variations in the throw of the fault. Where the throw is great, more recent beds are preserved beneath the surface: where it is slight, on the contrary, it is only the older rocks which present themselves. Of course, within the cover-rocks, the vertical variations of facies are frequent. Such differences in age in the rocks making up the surface are usually expressed as differences of resistance to erosion.

When, as frequently happens, uplift of an antéclise is renewed and the surface cutting across the fault is deformed, differential erosion is again instigated revealing basins coinciding with the outcrops of the weaker beds. Such a history has been described from the northern extremity of the western margin of the Morvan (see the geological maps, *Avallon* and *Chateau-Chinon*, 1 : 80 000 scale).

All that is needed here is a few examples. The more complex the combination of factors, the greater is the resultant variety. Thus, one may pass gradually from the type example to the individual cases.

Conclusions

Platforms and geosynclines constitute two structural macrotypes which must be compared in outline. Obviously the differences most evident at first sight are:

1. Differences in the rhythm and speed of deformation. Geosynclines are normally affected by movements which are more rapid, severe and, above all, more variable so that we always link them with instability. We have examined the relationship between this character and seismicity, that revealer of deeprooted disequilibrium. Geosynclines appear due to tensions located in the lower part of the earth's crust. The platforms, on the contrary, are relatively stable regions and show a particular tendency towards gradual stabilisation. In fact, the older ones are the most stable: the Hercynian massifs are a transitional stage between geosynclines at their last gasp and the Pre-Cambrian shields.

2. Differences in the tectonic style. The compartmentalisation of geosynclines into relatively tiny wedges, subdivided in their turn by secondary features, contrasts with the vaster and, at the same time, more durable units of the platforms. But there again, a transition may always be found, passing from the final stages of geosynclinal development with a gradual consolidation augmenting the surface of the tectonically mobile elements,

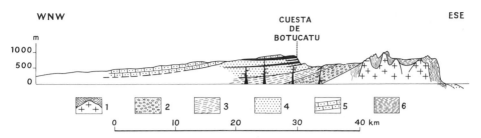

FIG. 3.18. The cuesta of Botucatu, in Sao Paulo state, Brazil (*after Ab'saber, 1956, p. 12*)
1. Pre-Cambrian granite and gneiss
2. Carboniferous conglomerates, sandstones and tillites
3. Permian pellitic slates
4. Triassic Botucatu sandstone
5. Cretaceous sandstone
6. Pliocene basalt of S. Paulo
In black, basalt intrusions (dykes and sills in the Trias). The basement forms a
backbone, plunging in a series of faults to the east, and with a Tertiary basin. The
cover-rocks produce a peripheral depression and then a series of scarps due mainly to the
reinforcement of the sandstones by basalt

to the recently (Hercynian) consolidated but still relatively fragmented platforms, then the most ancient, Pre-Cambrian platforms and, finally, the most vast members, the great shields or cratons where over some hundreds of millions of square kilometres may be found the same palaeogeographical traits and the same evolutionary sequences. This platform–geosyncline contrast must not be regarded as a onesided view, for it is essentially dialectical in nature. The differences express themselves particularly in an evolution, the stages of which are of uneven duration, the lengths of the stages increasing as development proceeds. The phase of gradual stabilisation in the geosyncline lasts at least 50 to 70 million years, that is, as long as the two preceding phases taken together. The gradual stabilisation of the platforms demands even more time. In fact, we may distinguish tectonic regions in terms of their latest period of folding: 210 to 260 million years ago (Hercynian regions), 330 to 440 million years ago (Caledonian regions) and greater than 550 to 600 million years ago (Pre-Cambrian regions). A deceleration in the rate of tectonic action is produced in time, with transition from geosyncline to platform. Things change gradually yet become more and more different: this is the very denial of the cycle concept, which supposes a return to an earlier condition. It can be seen, therefore, that Davisian concepts provide an extremely false idea of the very nature of such things. They can only mislead research.

This general kind of evolution is not necessarily linear, leading to one final result. It involves interruptions, even reversals which, however, never follow exactly along the lines already traversed. The gradual stabilisation of the platforms can happen again as a result of particularly violent and intense stresses which rift them and produce the rift valley systems. This

207

process can continue as far as the appearance of regenerated geosynclines. Although, because of their nature, they are particularly subject to vertical movements, the platforms may also be affected, in certain types of earth movements whose true causes still escape us, by tangential stresses in the same way as the neighbouring geosynclines. Such compression is responsible for the *plis de couverture*.

It follows that apparent diversity of structure takes on a profound order controlled by the very evolution of the Earth. It is this which provides us with the main thread, the framework of that dialectic opposition between internal and external forces whose result is morphogenesis.

4
Faults and volcanoes

It remains to consider the two types of manifestation of the internal forces: faults and volcanoes. They have been grouped together in this chapter for three reasons:

1. They are more localised phenomena than geosynclines and stable blocks. Volcanoes are often found in belts, it is true, but such belts are never as long or as wide as geosynclines. The largest accumulations of volcanic outpourings, as in the Columbia–Snake basin of northwestern United States, cover less than 2 million sq km and are of much less extent than the old shields. Faults usually bound regional structures that are parts of much greater morphological units. Such is certainly the case with the graben and horsts of hercynian Europe, and the same is true of the immense fracture zone of East Africa despite its large dimensions. When we move from geosynclines and stable shields to faults and volcanoes we are going down the dimensional scale; and since the plan of this book has been to lead from the greater to the smaller, the study of faults and volcanoes falls here into its proper place.

2. Faults and volcanoes are of smaller dimensions than geosynclines and shields and are also subordinate to them on the world structural map. The different dimensions reflect dynamic differences. Faults are breaks in the earth's crust; they occur whenever the crust is subjected to violent stresses, and are not confined to any particular type of structure. They are found equally in geosynclinals and stable blocks. The subdivisions of geosynclinals are often bounded by faults, initiated during the early stages of foundering and confirmed by the movement of blocks during the later phases. Commonly made rigid by the intrusion of granite, the stable block breaks more readily than it folds, and all the deformations encountered in an examination of such areas are of the fracture type. Volcanoes are also found on many different types of structure: they occur in the folded circum-Pacific arcs as well as on the stable block of East Africa. Faults and volcanoes are thus manifestations of a minor character that intrude into all kinds of major structures.

3. Faults and volcanoes are also often associated. All faults do not permit the outpouring of lava; but nevertheless the major fault zones are generally

studded with volcanoes, whether it be along the cordilleran fold and fault belts, the East African rift valley system or the much smaller faults of the Auvergne. The most important faults, those that affect the whole thickness of the earth's crust, do facilitate the rise of molten magma; they are also usually associated with gravity anomalies that assist its mobility. Though all faults are not accompanied by vulcanicity, all volcanic regions are faulted.

We therefore first study faults and then vulcanism.

Faults and fractures

It is impossible to understand the influence of faults and fractures on relief without knowing how they are formed.

Part played by faults and fractures in tectogenesis

THE TECTODYNAMICS OF FAULTS

Faults manifest themselves in the shearing action that results from a rupture of the rocks which form the earth's crust. They can thus be studied from two viewpoints, mechanical and evolutionary.

Mechanical aspect

The shearing action can only be produced in a solid rock, and it takes place when the rock is subjected to stresses beyond its limit of resistance. To understand the formation of faults it is thus necessary to take into consideration the mechanical properties of the rock on the one hand, and the nature of the stresses on the other.

Lithology is of great importance in faulting, for on it depends the capacity of the rock to resist stresses. The most rigid and inelastic rocks are those that are least able to accommodate the stresses; they break easily and develop numerous fractures. This is the case, for example, with quartzites and strongly cemented sandstones; they will not bend, they just break. The plateaus of Ordovician sandstone in western Sudan, for example, although part of a stable block, are riddled with cracks in various directions which cut across one another. The cracks bear no trace of shearing on their sides, and there is no displacement; they are not faults, but they are of tectonic origin and result from fracturing. They play an important geomorphological role in guiding the weathering of the sandstone flags, not merely by facilitating the detachment of sandstone blocks but also by aiding the downcutting of streams along the fracture zones. Conversely, marls, clays and sands do not crack. Layers of sedimentary rock consisting largely of great thicknesses of these rocks are more likely to bend without fracturing, and are particularly apt for the development of cambering. Sand especially acts like a mattress in smothering fractures.

The geometry of the beds also plays a part. Thinly bedded strata are the most supple. Each stratum can move slightly in relation to the one above, and this reduces the possibility of fracture. This bed-by-bed movement is frequent in folds, and is particularly important in thinly bedded limestones, in which fractures are much rarer than in massive and thickly bedded limestones or in consolidated conglomerates. On the contrary, a strong rock, if it is well jointed, may behave, in a massif, as a plastic rock, for slight movement can take place along the joint planes which renders the whole rock mass capable of being bent. This is the case with the Peruvian granite batholith, caught up in the intense Andean folding. It has taken on a domed structure, and has even bent without breaking, thanks to the innumerable very slight movements that have taken place along the joint planes which cut up the granite mass into a gigantic pile of cuboid blocks.

We must not therefore, overemphasise the contrast between bending and fracturing. There is a gradation from the one to the other, depending on the nature of the rock and the degree of tectonic force exerted. Rocks that are sufficiently supple will respond to a gentle and localised stress by flexuring; subject them to greater stress and they will break: the fold becomes a fault. This kind of process can happen in the course of time without any paroxysmal tectonic effort. And it can also be seen spatially. A deformation may appear first as a simple flexure, then as a monocline and finally, as the difference in level on the two sides increases, as a fault. The transition from fault to fold may also take place vertically (Fig. 4.1). In the Paris region, underground works have shown that faults in the Bartonian formation die out upwards and are replaced by folds of smaller amplitude in the overlying gypseous marls of Ludian age. Often, in the sedimentary rocks that overlie a stable block, faults die out upwards, passing into folds which themselves decrease in intensity in a vertical direction. This is why the progressive denudation of the sedimentary cover,

FIG. 4.1. A fault passes upwards into a flexure
The alternation of pliable and rigid strata produces the following effects: (i) a diminution in the throw of the dislocation, (ii) the conversion of the fault into a flexure, and (iii) the subdivision of the fault into several small fractures

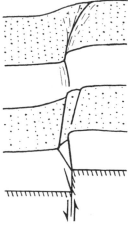

exposing deeper and deeper beds, commonly brings to light tectonic features ever more and more broken and more intense. Such disharmony is not confined to the folded mountain chains, it is only more accentuated there by reason of the greater intensity of the tectonics. The thick accumulations of soft sediments in foundered basins usually conceal fractures which increase in number and in intensity with depth.

Fɪɢ. 4.2   Tectonic breccia at Delphi in Greece
Jurassic limestone crushed and recrystallised

When the strata succession consists of a heterogeneous mixture of rock types, the mechanical properties of the different rocks affect the faults, which may be curved or even zigzag. The fracture meets resistance from a particularly massive rock, and curves or angles round it. It sometimes happens that a fault becomes subdivided into several fractures which intermingle in rocks that shear easily, and then resumes its single character on reaching a bed that offers more resistance to shearing. This is why a fault is rarely the single cleancut break that, by reason of scale, appears on geological maps. Almost always it comprises a sheaf of fractures, all with more or less the same direction and the same hade, fractures that meet, separate, bifurcate and rejoin or lose their identity in an entanglement the complexity of which varies with the magnitude of the fault. Naturally, the 'sheaf' effect is modified in relation to the properties of the rocks and varies from one stratum to another. In general, in the more resistant rocks the fractures that make up a fault are fewer, larger and more cleancut, and slickensiding, with the polishing and striation of the adjacent surfaces, is more pronounced.

The time aspect

For fracturing to take place, there must be stresses of sufficient intensity. The nature of the forces being brought into play implies a long period of preparation followed by a sudden release. That is why faults are commonly associated with earthquakes. Within the framework of regional tectonic evolution, the stresses are either tensional or compressive. Building up gradually, they eventually reach the limit of resistance and the rock breaks. An abrupt movement takes place which relieves the stress and restores equilibrium. This movement causes the shuddering that is an earthquake; it results in the relative displacement of the rocks on either side of the fracture.

The association of faulting and earthquakes is a continuing phenomenon. Almost all the major earthquakes occur in faulted areas. It has been possible to identify their epicentres along known lines of faulting—as in the

FIG. 4.3. Slickenside near Corveissiat in the Jura
On the left, crumpling of thin limestone beds in contact with the fault; in the centre, the curved and polished slickenside; on the right, tectonic breccia

case of the sub-Atlas fault that produced the Agadir earthquake, in the case of the Skopje disaster of 1963, and in association with the San Andreas fault that caused the San Francisco earthquake of 1906. It often happens, as Japanese seismologists have discovered, that the phenomena have a certain oscillatory character. During the preparatory period, slow deformations are produced, as for example the raising of block A and the depression of block B: then, during the earthquake, abrupt movements in the opposite sense take place, with block B being raised and block A lowered. This is easily understood. Before the earthquake the stresses encounter resistance which tends to hinder them; a limited movement is produced, but when the resistance is finally overcome things go too far and a kind of rebound occurs. The limited elasticity of the crust no doubt contributes to this effect.

FIG. 4.4. Slickenside in basaltic breccias, north of Chirilagua in Salvador
    Broad polished surface, conducive to water erosion and weathering, and lending itself to scouring by land-slipping, as occurred here before the construction of the road

Such oscillatory movements are frequent; they may be compared with the pulsations that have been observed in stable areas by repeated accurate levelling: an area may rise for a while, then subside and rise again. It is of course difficult to measure the rate of movement of faults, since we can only deal in very short periods of time, very different from the scale of geological time. What matters, geomorphologically, is the end product of the movement, and this may very well be different from that indicated by short-period measurements.

Faults have a tendency to recur, so that existing fault lines are particularly prone to renewed movement, and for this to take place, a shearing stress much smaller than the one that caused the original fracture may suffice. All that is necessary is for the forces that produce the relative movement of the two sides of the fault to be great enough to overcome the friction along the fault plane. This is relatively common, and is much easier to achieve if the two sides have a tendency to move away from each other (tensional faults), for the friction is then minimal. On the contrary, when the two sides tend to approach one another, as in compressional faults, friction is at a maximum. This is what happens with thrusts and reversed faults. Earthquake shocks occur when the frictional resistance is suddenly overcome.

We can thus understand why faults and earthquakes are not always associated in the same way. Movement may sometimes occur freely along a tensional fault because there is no great resistance; earthquake shocks will be feeble or non-existent. In Central Asia, according to Rantsman, movements that occur rapidly and uniformly seldom cause earthquakes. On the

FIG. 4.5. Gaping faults in micaschist, in the Cordillera de la Costa near Valencia (Venezuela)

215

contrary, fractures subjected to displacements that vary rapidly over a short distance are generally the source of sharp earthquakes that alternate with periods of slow and slight movement. The occurrence of earthquakes, if not due to volcanic action, is an indication of movement along fault lines. Such movement, however, may occur without much seismic activity, as would appear to be the case in Alsace, where there was important faulting in Quaternary times but no earthquakes at the present time; but then the Alsatian graben is characterised by tensional faults.

It is thus not surprising that movement along fault lines, like other tectonic movements, should be extremely variable, both in time and in space. The faults that generate strong earthquakes are not necessarily those with the quickest rate of displacement. The maximum actual surface displacements recorded during an earthquake are of the order of 2 or 3 m, as in the case of the San Andreas fault in the disastrous San Francisco earthquake. In New Zealand an earthquake in 1855 in the Hutt Valley area tilted a block 50 km long, raising one side 3 m above the other. In Chile, the Concepcion earthquake of 1960 raised some areas by 1·5 to 2 m and lowered others 0·5 to 1 m. In Japan, the Fukui earthquake of 1948 caused an horizontal displacement of 1·67 m and a vertical one of 0·75 m. In the Neo valley shock of 1891 there were horizontal movements of 1·6 to 2 m and vertical ones of 0·6–0·65 m. On the other hand, in Hungary, without any notable shocks or catastrophes, movements of 2–3 mm a year have been recorded for some fifty years. Since catastrophic earthquakes only occur at intervals of a century or two in the regions mentioned above, it is clear that gentle movements can produce equally important deformations over thousands of years. Such differences in tectodynamics are only of interest to geologists; but earthquakes are also the cause of geomorphological phenomena such as fault scarps, as we shall see.

THE DISPOSITION OF FAULTS

The disposition of faults is conditioned by the distribution of tectonic stresses on the one hand, and the heterogeneous nature of the earth's crust, on the other. We can adopt two approaches to the subject, one statistical and quantitative and the other qualitative, relating faults to regional structure.

Quantitative analysis

Cailleux has made most progress along this line, taking California as his example. His results have been confirmed by similar analyses in other folded areas, notably the Liège Carboniferous basin. From these studies it appears that faults form coherent systems and are not just distributed at random. Their various characteristics have statistical relationships, rather like those which have been worked out for hydrographic systems.[1]

The laws deduced from statistical analyses are as follows:

1. The average distance apart of faults is very variable. But there is a lower limit, which may fall to about 250 m, which depends on the nature of the beds and the importance of the faults. On the Alsatian border of the Vosges, the spacing of faults with a throw of several dozen metres, in the Mesozoic rocks, is between 600 m and 1 km.

2. The inclination of faults also shows a modal distribution. In California the mode is 67–70°. Contrary to what one might suppose from indications commonly shown on geological sections, faults are seldom vertical, and an inclination of 70° is much the most frequent. This has two important consequences: first, that fault scarps can rarely be subvertical; however, fault planes usually slope at a much greater angle than the 45° which is the angle of repose for screes, so that the removal of rock-masses from the face of an active fault scarp is easy. Secondly, faults rarely have only vertical displacement; more commonly there is a horizontal as well as a vertical component, as we have already indicated above. The lateral displacement along sloping faults often causes a compression of the beds on one side or the other, and strata are often slightly folded or even overturned in the vicinity of faults.

3. The length of faults, like their spacing, varies from region to region, depending on the tectonic stresses and the lithology of the rocks. In California, it is commonly between 300 and 1 000 m.

4. There is a correlation between the length of faults (L) and their throw (D) (expressed of course in the same units, metres or kilometres), which can be expressed by $D = kL^a$, a formula that recurs in studies of hydrographic systems. The coefficient k and the index a have regional values that reflect the mechanics of the fracturing in each structural unit. Cailleux's graphs, however, show a very wide dispersal. Nevertheless, the general conclusion is that the longest faults also have the greatest throw, and vice versa. This is certainly the case in Alsace, but it seems less so in Lorraine, on the plateaus of Bajocian limestone, where some faults of quite small throw can be traced for long distances. And it is certainly true of the faults that are involved in presentday earthquakes.

5. An expression of the frictional resistance encountered during movement along a fault is given by $L \times D$, which corresponds to the surface of the

---

[1] We should be quite clear as to what such statistics can show: they indicate that a certain characteristic is *usually* in a certain relation to another characteristic. They show what is the most frequent, or 'normal', arrangement. But this arrangement is not the only one and many individual cases will deviate from it, though the number of deviant cases will be smaller as the degree of deviation gets greater. Statistical correlations define *types* and enable us to understand them without hindering our ability to study the characteristics of each individual case. Statistical analysis thus constitutes a valuable aid in geomorphological classification.

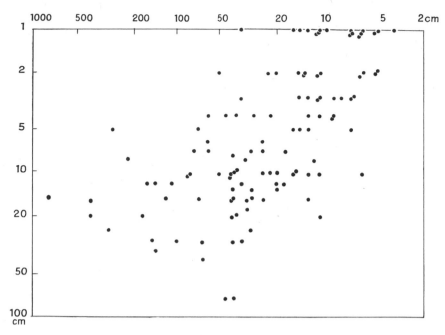

Fɪɢ. 4.6. Relationship between the width of a fault and its throw (*after Cailleux, 1960, p. 109*)

Vertical axis, the width of the fault, i.e. of the fissure or of the breccia that has filled it. Horizontal axis, throw of fault. A strong correlation is apparent. The width of faults increases with their throw. Naturally, the wider the fault-zone, the easier it is for erosion to take place along the fault-plane

frictional plane. Since *L* and *D* increase simultaneously, it is obvious that their product grows very rapidly. It is the large faults that dissipate most energy. According to Cailleux, some 2 to 20 per cent of all faults represent 50 to 60 per cent of all the energy expended.

Faults and regional structure

In order to place faults in their regional and structural context, their relationship to the dip of the rocks must be defined, and also their relationship to each other. The relations between faults and dip may be defined as follows:

1. *normal faults* are those with a steep inclination (or small hade), in which the beds on the downthrown side rest in part upon the fault plane;
2. *reversed faults* are those in which the beds on the downthrown side are partly covered by the fault plane and by the upthrown beds which overlie it; in other words, the hade is towards the upthrown side, which is thrust over the downthrown side;
3. *conformable faults* (*failles conformes*) throw down the beds in the direction of dip;

218

4. *contrary faults (failles contraires)* raise the beds in the direction of dip, thus giving any individual stratum a sawtoothed appearance in a cross-section.

Faults are just as much associated with folded structures as with tabular ones. In the case of folded structures, as we have seen in Chapter 2, several varieties can be distinguished. Faults that result from tears produced during the formation of the folds end up as fold faults, fault splinters, and *nappes de charriage*. Faults which existed before the folding took place, and were thus deformed during the folding, produce fold faults, anchorage and pinch structures, etc. Finally, faults occurring after the folding cut the folded beds and give faulted folds.

The formation of fault folds usually results from compressional movements; tear faults occurring in folds are sometimes the result of tension, whilst faulted folds are usually the consequence of tensional stresses. This explains why the different forms appear at different times during the evolution of folded mountain chains, in response to the tangential displacements that generate the folding, which alternate between phases of compression and tension.

In tabular structures, fractures are usually major tectonic accidents which result from sudden stresses. They initiate the individualisation of compartments during the appearance of regenerated geosynclines. They do not have to come to terms with the dips produced by other tectonic mechanisms and it is they which usually control the dips because of the tangential movement they engender by their obliquity. Very often in more or less horizontal strata, the approach to a fault is marked by upturned beds with abnormal dips—$10°$ for example in an area where the normal dip is only $1°$. The uplifted blocks close to the faults show dips which are intermediate between those of tabular areas (under $3°$) and those of folded regions (generally more than $30°$). But such high dips occupy but small areas, which explains their feeble representation in statistics relating to large areas. Nevertheless, such local departures from the normal may have considerable geomorphological importance as in the case of short but pronounced monoclines or through their influence on slope formation and the development of asymmetric valleys.

The relationship of faults to each other is of especial interest in tabular regions and in the collapsed arch structures in folded mountain ranges (forming intermontane basins).

Most of the rift valley structures which result from collapse between parallel faults are to be found in regions subjected to tensional stresses which have tended to pull apart the edges. The origin of this distension of the earth's crust may be found in an uplift due to a local swelling. This would appear to be the case in certain mountain systems on the edges of continents, such as the Andes of Chile and southern Peru. The alignment of the intermontane basins in the Central Valley of Chile marks out a foundered zone at the contact of the much uplifted cordillera of the Andes

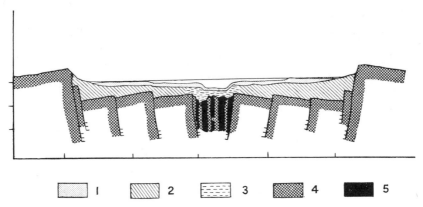

Fig. 4.7. Section across the northern part of the Red Sea, from geophysical data (*after Drake and Girdler, 1964, p. 489*)
1. Recent sediments and coral reefs
2. Limestones and evaporites
3. Sediments interstratified with volcanic rocks
4. Basement
5. Basic intrusions

and the coastal ranges, which are much less uplifted but which border the oceanic depths. The distension may also be due to a tear in the earth's crust and the stretching apart of its edges. It is to this kind of mechanism that the great rift valley system of East Africa has been ascribed. Narrow foundered strips are there framed by uplifted blocks that overlook them in a series of faulted escarpments. The foundered areas are more or less depressed; they come and go, bifurcate, disappear, rather like the tears in an old cloth that is pulled apart. Such rift valley systems are interpreted by some geologists and geophysicists as evidence of a tendency for the globe to expand and increase its volume.

However this may be, rift valley systems are usually characterised by three things:

1. Uplifted edges which tilt outwards—as in the case of the Vosges and Black Forest on either side of the middle Rhine graben. These bordering

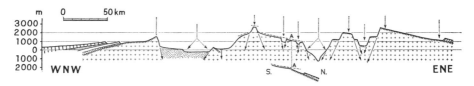

Fig. 4.8. Section across the faulted zone of the northern Red Sea (*from Vaumas, 1961*)
 The tabular cover of Egypt has disappeared by erosion from the Arabian upland.
In the faulted zone the basement is broken into blocks, irregularly raised, lowered and tilted

blocks are nearly always very rigid, with crystalline rocks at the surface or at the shallow depths (like the pre-Cambrian in East Africa and the hercynian granite beneath Triassic sandstone in the middle Rhine). Sometimes the graben coincide with less rigid rocks, like the slightly altered palaeozoics of Alsace.

2. The stepfaulting of narrow strips on the edges of the graben, showing how the whole process occurred. Many of these strips have been tilted and subjected to rotational movement rather like that which occurs in landslides. They have, as it were, been drawn into the void of the graben. In Alsace, for example, certain thin slivers of Jurassic limestone at the foot of the Vosges, near Colmar, show dips of as much as 45°. Tilts of 20° are common on the borders of rift valleys. There is little slickensiding of the faults, showing that friction was at a minimum—except of course, in the case of reversed faults. The beds are often stretched, fissured with open chasms, sometimes almost literally disjointed, as in collapsed structures.

3. Gravity anomalies indicate a rise of magma beneath the graben, perhaps facilitated by the decompression due to the stretching of the upper layers of the crust. Both the stepfaulting and tilting of the sides and the rise of magma seem to combine to fill the void created by the collapse of the arch. The opening of the cracks facilitates the emission of volcanic lava, and this is why, in general, rift valleys are accompanied by volcanoes, as in East Africa and the Central Valley of Chile.

It is clear that faulting has most influence on relief in those areas where the structure is one of blocks outlined by major faults. Such areas are residual geosynclines, regenerated geosynclines and rift valley systems.

Geomorphological role of faults and fractures

Faults and fractures influence relief in three different ways: by the tectonic displacements that they create, by the modification of the rocks in their vicinity (crush zones, breccias and mineral veins) and by the juxtaposition of rocks of different weathering qualities along the line of the fault. These repercussions also have a time relationship with the tectogenesis. Tectonic relief appears during the actual earth movements and remains for a variable time in the form of residual relief, the longevity of which depends on the resistance of the rocks affected by the fault. Similarly, lithological modifications in the vicinity of the fault may also exert an influence during the formation of the fault and for a long time afterwards, especially in the case of quartz veins, for example. On the other hand, differential erosion takes place mainly after the cessation of movement, in beds which are tectonically stable.

TECTONIC RELIEF FORMS

These are the direct result of the relative displacement of the rocks by the fault, creating a fault scarp which corresponds theoretically to the throw.

Faults and volcanoes

Such features are commonly called fault scarps to distinguish them from faultline scarps, which are the result of differential erosion and are therefore influenced by lithological as well as tectonic factors.

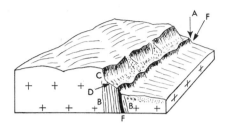

Fig. 4.9. Original fault-scarp, on the side of the Waitaki valley in New Zealand (*drawn by Schneider from a photograph by Cotton*)
F. Fault
A. Small slickensides
B. Crush zone
C. Weathering crag
D. Scree derived from C

A true fault scarp appears at the same time as the fault itself, and it grows with further movement along the faultline. It is thus an active tectonic form. By definition, all active fault scarps are original, so it is not necessary to add the adjective 'active'. But when faults are sufficiently active and affect resistant rocks, the scarp persists for some time after the movement has ceased, and the slope is only gradually reduced. We thus have residual fault scarps, that are relics of the stresses that caused the original fault.

The shaping of active fault scarps

Movement along the faultline results in the development of a talus scree which is subjected to the normal processes of slope formation. But these processes are influenced by two consequences of the actual faulting.

In the first place, since rocks are normally jointed, the fault does not sever them quite as cleanly as a knife going through butter; moreover we have already noted that the faults themselves are usually sheaflike, splitting and ramifying. In most cases the rock is cut by more or less parallel fractures that divide and rejoin. These fractures guide the erosional processes by facilitating the penetration of water, which permits chemical as well as frost action; and they may detach whole rock masses which are rendered unstable and liable to roll down the slope. If solid rocks are cleancut by a fault plane several score or even several hundred metres in height, the actual fault plane will appear directly in the relief, like a wall, inclined as faults usually are. But this would presuppose great lithological homogeneity. When rocks of different mechanical properties are involved, the fault will split in the more friable beds and resume as a single fracture in the more rigid rocks. In the latter case, the more intense contact of the sides of the fault plane has caused slickensided surfaces, polished and fluted, and often covered with a coating of crystalline quartz, calcite or other minerals. Because of the modifications of rock structure brought about by the faulting, fault scarps seldom present a solid wall coinciding with the fault plane, and they are generally irregular, with broken zones with rock pinnacles and a chaotic mass of boulders. Occasional sections of the fault plane itself, hardened by the friction, or 'armour plated' to translate literally the German term, give steep and smooth slopes alternating with the rough boulderstrewn screes. These are usually most numerous in the middle part of the scarp, for lower down they are obscured by debris, whilst above, the whole of the faulted zone having been converted into fallen blocks, the scarp face is formed by the solid rock behind the fault plane. It often happens that small oblique faults branch from the main fracture giving narrow zones of crushed strata littered with broken boulders, or making nicks in the scarp edge. In cold climates, nivation can make use of such belts.

Secondly, movement along the fault is accompanied by earthquake shocks. The effect of these is particularly important when the rocks are already broken, or where piles of detached blocks are in unstable positions. During earthquakes, rock falls commonly occur on fault scarps. In massive rocks, blocks of as much as a hundred cubic metres may roll down as far as the first obstacle or right down to the bottom of the scarp. Screes of smaller material start to move downhill, and the zones of broken rock are swept clean of all the masses that have become detached from the rock face. When there are beds of marl or clay in the scarp face, the shocks may

FIG. 4.10. Fault-scarp at the junction of the Cordillera de la Costa and the Caracas basin, in Venezuela

Gneissic rocks (originating in a granitisation of Cretaceous age) and Quaternary tectonics that are still operative. The line of fracture is represented by a valley. On the left (the upthrown side) part of the Cordillera de la Costa, with the original fault-scarp inclined at an angle of 45° and notched only by short but deep, parallel ravines

223

Fig. 4.11. An original fault-scarp near Lagunillas, in the Venezuelan Andes

The fault-scarp affects the older Quaternary formation, that outcrops on the shoulder. It is but little dissected, save by a few gorges; concentrated stream-development has not yet taken place

generate mud slides of waterlogged material that may undermine the harder beds that lie immediately above the clays. Volcanic formations, with alternations of fine ash and solid lava, are particularly liable to such occurrences, as has been shown in Peru and Chile. Earthquakes do not only affect active fault scarps, they influence all other slopes in the region. But on the one hand the shocks are most violent in the vicinity of the epicentre, which often coincides with a fault plane, and on the other hand the fragmentation of the rocks in the vicinity of the fault means that they are peculiarly susceptible to shocks by reason of their instability. Movement along faults accentuates the role of gravity, especially along the scarp itself. In very fragmented rocks, as often in the Andes, the whole scarp may become a scree with an average slope of 45° or more, interrupted only by rock stumps that stand up like ruins. Traces of the slickenside surfaces can often be found on detached and fallen blocks, but no part of the fault plane itself is visible.

The physical form of active fault scarps may thus vary from a steep wall formed by the fault plane in solid rocks to a slope littered with fallen blocks, crumbling and unstable. In the latter case, the slope is always displaced from the line of the fault and cuts across the crushed rocks in its vicinity. When one can locate it, the fault line is usually at the foot of this slope. But it is often obscured by the fallen boulders or by the alluvial material that spreads out from the foot of the scarp, especially in dry regions and on the edges of rift valleys. When rivers cross the scarp, the outcrop of the fault can often be traced in their beds.

Active faults and the hydrographic system

The movement of faults causes disturbance of the hydrographic system. Indeed, the movements are often so rapid that the uplift of a block takes place at a greater rate than that of the downcutting of a stream in solid rocks. Furthermore, faults are often accompanied by the warping and tilting of blocks in a way that is not necessarily in conformity with the existing drainage system. We may examine in more detail each of these problems.

The erosion of an active fault scarp depends on its position in relation to the original drainage system. On the edge of rift valleys, streams commonly cross the boundary faults. The foundering generally takes place before being accelerated by the faulting. Moreover, the faulting is not always localised in the same alignment throughout the period of movement. Sometimes the faulting moves inwards, which restricts the size of the graben; sometimes it moves outwards thus enlarging the graben at the expense of the neighbouring horsts. For example, in Alsace to the west of Strasbourg, the neighbourhood of Achenheim forms a block that foundered in the early Quaternary and was covered by the alluvia of the Bruche and Rhine; later it resisted further subsidence and was slowly raised, thus permitting the accumulation of thick loess deposits on the dissected lower

FIG. 4.12. A faulted coastline, with tilted raised beaches, Baring Head, New Zealand (*after Cotton, 1951, p. 108*)

A staircase of raised beaches on the emergent block. The coast corresponds to the fault-scarp. The emergent block is tilted along the fault line, as can be seen from the way in which the base of the dead cliff (above the lowest raised beach) rises towards the foreground. It actually rises from 90 m above sea level in the distance to 160 m on the edge of the valley

Quaternary surface. On the contrary, near Sélestat, downfaulting moved in the direction of the Vosges during the Quaternary, thus allowing the deposits of the Würm glaciation to accumulate on the ancient rocks. Certain active fault scarps coincide with water partings: but others are crossed by pre-existing rivers.

When the fault scarp forms a water parting, the evenness of the scree slope is not conducive to the formation of streams. From place to place, however, little hollows will appear at the top of the scarp, which in the nature of things has been longest exposed to erosion, and in humid climates these will form little drainage basins from which steep ravines will lead down the slope. The cross-profiles of these gulleys cannot become smoother because of the continued downcutting due to the movement of the fault, and their lower part forms a narrow gorge. Sometimes, indeed, when the rocks are difficult to erode, the gorges are suspended, and the streams cross the faultline in a waterfall; the slope is thus but little dissected and the streams are scarcely incised at all into the rock. Between the streams there are large stretches of the scarp aligned more or less along the fault. This state of affairs is usually called dissection in facets (Figs. 4.15 and 4.16).

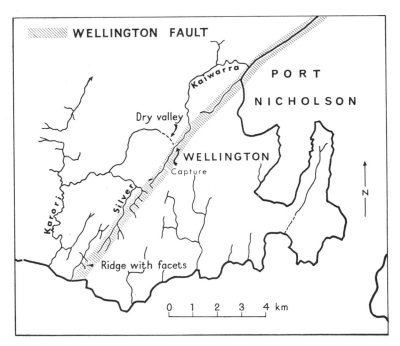

FIG. 4.13. Influence of a fault on the river pattern, at Wellington, New Zealand *(after Cotton, 1951, p. 64)*
   Streams are aligned along the fault and the zone of crushed rock that accompanies it, as a result of captures. Several right-angled elbow bends where the streams encounter the fault zone

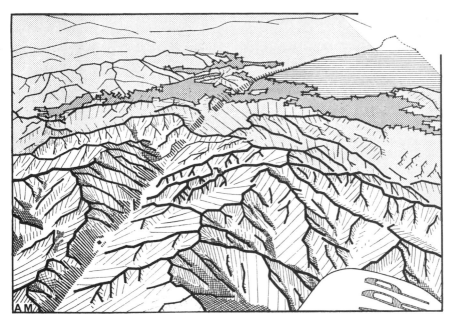

FIG. 4.14. The Wellington fault *(drawn from a photograph in Cotton, 1957, by A. Maurer)*
The fault creates a very sharp escarpment, indicated in the drawing by parallel hachures. It can be followed in an arc from the left-hand bottom corner to the right-hand top corner. It cuts across a landscape of varied relief and dissection

At one time it was thought that these facets were a peculiar characteristic of original fault scarps. But this is not the case; the facets remain, for quite a long time, along the face of residual fault scarps. Moreover, many active fault scarps do not have these facets, which are in fact a function of the lithology. They can only develop if the scarp coincides, at least in part, with the fault plane, and this requires a sharp and clean fracture, which can only occur in certain types of rock.

When active faults cut across an existing hydrographic system, their movement interrupts the smooth evolution of the rivers. Depressions become downfaulted at a more rapid rate than they can be filled with alluvium, and lakes result. Other lakes may form in valleys that are truncated by a fault that raises the land on their lower course. This sort of thing happens whenever the downcutting of the river is slower than the uplift due to the fault. In the case of uplifts amounting to several centimetres a century, this will always happen if the rocks are hard and not easily eroded. In softer or more fissured rocks the river can incise itself into a gorge. For example, in New Zealand, the Hutt river, after the 1855 earthquake, had to cut down 2–3 metres, and as a result ceased to be navigable, because of the tilting of the area through which it flowed.

Horizontal movements can also affect the drainage system, as Cotton showed in New Zealand in 1957. Valleys are shifted laterally along the tear fault, giving the river a zigzag course. This lateral displacement may actually block the valley like a sliding door, and cause the formation of a lake which eventually empties itself by eroding the crushed zone along the faultline. Such movements are always accompanied by minor undulations of the land on one side of the fault or the other, and these may also cause obstructions that interfere with the drainage pattern. The dissection is often different on either side of the fault because rocks of different qualities are brought into juxtaposition. And branches of the main fault frequently localise tributary valleys, for stream erosion has a strong tendency to coincide with the zone of tearfaulting, and this is accentuated if the fault is active.

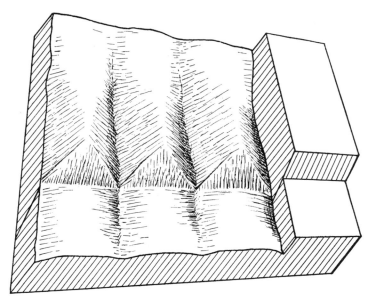

Fig. 4.15. Median facets on ridges marking the position of a fault (*after Cotton, 1950, p. 227*)
The fault cuts across a relief made up of parallel ridges. Along the valleys there is no interruption in the long-profile of the streams, for the rocks are incised without difficulty. But the original fault-scarp is apparent on the ridges in the form of facets

Faults are also the cause of river captures and drainage reversals. In California, Pouquet has described the capture of Furnace Creek, a longitudinal stream flowing on the backside of a tilted block, by the Gower Gulch, a very active ravine on the scarp slope, which has truncated the pre-existing longitudinal stream. In East Africa, on the edges of the great lakes, however, it appears that streams originally flowing towards the lakes have been turned aside and even reversed, for the blocks that they cross have been tilted away from the lakes. In a tropical climate and with granite

Fig. 4.16. Fault-scarp cut up into facets, Deep Springs Valley, California
(*drawn by Schneider from a photograph by E. Blackwelder*)
   Granite rocks and an original fault-scarp

rocks, vertical downcutting has not been able to keep pace with the uplift
and the runoff has been reoriented in conformity with the tilt; it follows the
new slope of the granite blocks instead of continuing in the opposite direction. Modifications of a similar character have been described on the edges
of the Fossa Magna in Japan.

Residual and composite fault scarps

Movement along faultlines often continues for a very long time. In the great
rift valley system of East Africa, for example, it has been active since the
early Miocene. Within the regional tectonic framework not all the faults
are active at the same time. A certain fault may alternate, during one period
of tectogenesis, between periods of movement and periods of quiescence.
For example, the fault that bounds the Vosges to the south of Saverne was
active during the Villafranchian, and again in the middle Pleistocene
(during and after the Mindel-Riss interglacial). Faults also work in relays
as it were, both in time and in space. Compensatory movements take place
within the whole structural framework of a region. Normally faults will
behave in a jerky or staccato fashion, with sharp movements alternating
with periods of slack or no movement. During the active periods, the scarps
made by the faults function as true fault scarps, whilst during the quiet
periods erosion converts these into residual scarps. After a time renewed
activity may occur, and if each of these phases is of sufficiently long duration the effects may be visible on the relief, in the form of a composite fault
scarp. A composite fault scarp is thus a scarp due to a fault that has been
intermittently active, so that the forms of erosion have varied between
those associated with active fault scarps and those of residual fault scarps.
We shall return to these after having examined residual fault scarps.

229

Fig. 4.17. Fault-scarps in Alsace at Gueberschwihr
On the right, an original fault-scarp of Quaternary age formed by the
Vosgian sandstone, straight, steep and wooded. In the foreground, a
fault-line scarp. The slope is of Jurassic limestone, the village at its foot is on
marl. The scarp is less abrupt, more scalloped, and disappears towards the
south where dissection is less active

Fig. 4.18. A transverse fault with
formation of truncated spurs (*after
Cotton, 1950, p. 750*)
The transverse fault cuts both
ridges and valleys, partly
disorganising the drainage. In valley A,
a blocked valley results in the
formation of a lake; in valley B, a
linking gorge exploits the crush-zone
of the fault

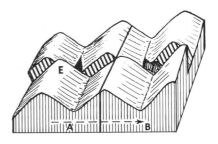

When movement along a faultline ceases, erosional forces are able to attack the scarp without having at the same time to counteract the tectonic forces. The fault scarp ceases to grow at its base, where the fault plane outcrops. It develops into an ordinary slope, subject to dissection that will vary in intensity with the angle of slope and the strength of the erosional forces; lithology becomes the most important determining factor, followed by tectonics. Escarpments formed from the alternations of hard and soft rocks due to a reversed fault will develop either as homoclines or as cuestas, depending on the hade of the fault. Some cuestas are thus derived from fault scarps. But morphoclimatic factors are involved also, because of the influence that they have on slope formation. The problem is thus a complex

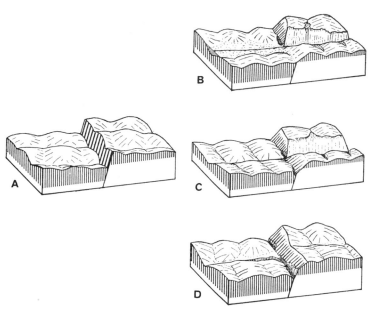

FIG. 4.19. Evolution of faulted blocks (*after Cotton, 1950, p. 739*)

A. The original form, as it would have existed on Davisian principles, without the interference of external with internal forces, that is to say, without the production of landslips and boulder-screes along the fault-plane during the earthquake accompanying the faulting.

B. The fault is crossed by a stream descending from the uplifted block. It incises itself into a gorge, cutting the fault-scarp into facets, and deposits alluvia on the downthrown block, the relief of which becomes subdued.

C. Similar to B, but in this case the regional circumstances entail the incision of the stream into the downthrown block as well. The dissection of the scarp is more emphatic, and small tributaries form valleys along the fault-line.

D. The rapid differential movement of the blocks and the general hydrographic pattern result in the latter being disrupted by the faulting. A reversal of drainage occurs on the upthrown block, with the main stream now flowing in the opposite direction, and a new stream develops along the line of the fault, causing a partial reversal of drainage on the downthrown block. In such circumstances, the fault-scarp may remain in evidence for a longer period.

one, and the variety is considerable. We can only suggest a few general points.

If the rocks are sufficiently resistant, the dissection of a faultline scarp will progress but slowly. Gradually small basins will be excavated in the slope, where stream-channelling is concentrated, and these will become larger and more ramifying as the main stream which they nourish cuts its way downwards and reduces its longitudinal gradient. There is a gradual retreat of the scarp wall from the line of the fault, and the facets, if any,

FIG. 4.20. An original fault-scarp, on the edge of the low-lying Waitaki valley in New Zealand (*drawn by Schneider from a photograph by Cotton*)

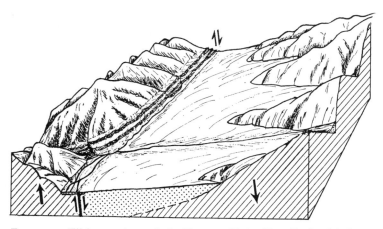

FIG. 4.21. Tilting against a fault, Hanmer Plain, New Zealand (*after Cotton, 1953, p. 216*)

The downthrown block is tilted towards the fault, and the movement is continuing intermittently. The results are as follows:

  (i) The gradual 'drowning' of the depressed and tilted block, with the fringes of the ridges disappearing under alluvial cones

 (ii) Asymmetrical accumulation of debris in the depression, with the alluvial cones, fed by the debris derived from the dissection that is itself increased by the tilting, coalescing on the depressed block

(iii) Elevated terraces along the fault-scarp, on the edge of the horst. These derive from a major river that was pushed up against the fault-scarp by the alluvial cones from the opposite side of its valley

become narrower. At the same time the slopes between the small valleys become degraded and retreat; they are reduced to a series of ridges, at first closely aligned and then becoming more and more staggered, with the narrower ones disappearing more quickly, whilst cols may open up in them, isolating *buttes-témoins*. As it evolves, the scarp loses both its character and its clarity. It becomes more and more irregular and sinuous, resembling any other sort of slope. This is why, in the Central Massif, many scarps of this kind, shapeless, blurred and imperfectly marked, were formerly considered, by Baulig for example, to be erosion scarps, that is steps linking two successive erosion surfaces. Petrographic analysis has shown, however, that in many cases they are old fault scarps, dating from the Oligocene or Miocene, which have not since been the scene of renewed faulting and so have been almost obliterated by the dissection. When the rocks are very resistant, residual fault scarps may survive long after the cessation of movement along the faultline, but progressively losing their distinctive characters. The original character is best preserved under an arid climate that permits pedimentation. Parallel slope retreat is very favourable to the maintenance of the scarp: but it is often difficult to identify it, for the 'glacis' that forms at its base soon levels off the surface, and the debris conceals the faultline. The actual scarp, still sharp but thus encumbered at its base, may be 1 or 2 kilometres or even more, from the fault, which has been levelled by the talus. It was this that caused the early geologists to underestimate the role of faulting in the development of pediments and the formation of basins in the arid southwest of the United States.

When a fault remains inactive for a sufficiently long time, its scarp begins to be dissected. Stream downcutting becomes more pronounced and the slope loses its rigidity. This is the case with the Vosges fault south of Saverne (Fig. 4.22). Here the conglomerate, though very resistant, was vigorously dissected during the Villafranchian period, and small foothills developed at the base of residual plateaus with cliffed edges. Chemical weathering, combined with intense stream erosion, has produced at the foot of the scarps a spread of quartz pebbles derived from the conglomerate, mixed with buff and orange-coloured sands. The erosion of the faultline scarp was quite pronounced during the second half of the Villafranchian and the older Pleistocene. Then renewed faulting occurred, probably at the beginning of the Riss or perhaps a little later. Small parallel faults affect the detrital cone of the Zorn that dates from the early Pleistocene. This renewed movement produced an outcrop, along the fault plane, of the friable sandstones that underlie the conglomerate. Despite the poor resistance of these rocks, the lower part of the scarp, corresponding to the recent movement, is not dissected at all. The little valleys etched in the conglomerate are suspended above a straight and quite undissected slope, covered with scree from which the finer materials have been washed out and redeposited at the base of the slope, on top of the slope deposits of the Villafranchian.

In general, the revival of a fault after a long period of repose results in the

233

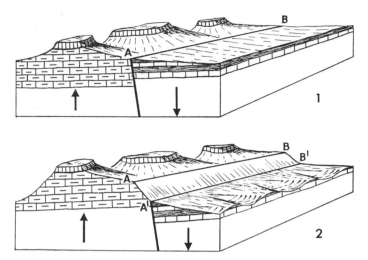

FIG. 4.22. Geomorphological result of renewed movement of a
fault, on the edge of the Vosges south of Saverne

1. The morphological stage reached in the Pliocene. The
uplifted block, formed of a hard conglomerate resting on Triassic
sandstones, is dissected into buttes with steep cliffed sides. The
fault AB is levelled by a concave slope that bites into the sandstone
above, and lower down, bevels the marls and limestones of the
Muschelkalk on the downthrown block. Quartz pebbles from the
conglomerate are scattered over this slope.

2. The present relief. The fault has been active again quite
recently (for Quaternary alluvia are affected). The dissected
surface of the uplifted block is perched above the original
fault-scarp (AA′–BB′) as a result of the further movement. The
valleys that crossed the slope are trenched and hanging; their
direction is sometimes reversed so that they flow behind the buttes.
The scarp has thus become a water-parting, and this enables it to
retain its freshness, with no notches. The dissection of the Pliocene
slope at its foot is being effected by small valleys whose heads have
not yet reached it

superimposition of the smoothly moulded relief, well dissected, of the upper
part of the scarp, upon the fresh features traced by the renewed faulting, in
the lower part. In other words, the upper part corresponds to the relief
form of a residual faultline scarp, whilst the lower part has the fresh and undissected form of an active fault scarp. Such composite scarps may of course
present many variations on this theme.

SHATTER BELTS AND DYKES

Faults are accompanied, as we have seen by shatter belts characterised by
many fractures oriented more or less parallel to the fault. The continuity
of the beds is thus interrupted. Sometimes this shatter belt guides the upwelling of magma in the form of a dyke. Through the medium of these

lithological changes, faults may continue to influence relief long after their activity has ceased, by reason of differential erosion. But the effect may work in either direction: erosion may be facilitated, or it may be hindered, according to the circumstances.

Shatter belts

Tensional faults, creating open fissures, facilitate erosion provided that there is no upwelling of magma. Some examples may be cited:

Shatter belts are often picked out by valleys of exceptionally straight alignment, the orientation of which is often in contrast with that of other branches of the same river system. Examples have been described in Portugal (Fig. 4.23) and in the southern part of the Bohemian massif. Many Norwegian fjords also correspond closely to fracture zones, which facilitated the movement of the glaciers and quickened their erosive action. When rivers pass over massive and impervious rocks like granite and quartzite, the existence of shatter belts facilitates the movement of underground water and gives rise to springs, so that chemical weathering is made easier. In tropical regions, such weathering may extend downwards to a depth of 100 m in fractured zones within crystalline rocks. In regions that are well dissected, a stream may follow the fractured zone, where it takes advantage of the loosened rock and has its volume increased from underground sources. Once it has formed this zone, it exploits it to the full by rapid headward erosion. But the fractured zone is generally quite narrow, and is often interrupted by massive rock that resists erosion. This is why streams aligned along such zones usually have but few tributaries and are just single trunks.

In limestones, fractures of all kinds play a most important role in deep karstification, and their effect is all the more marked if the rocks are thick and massive; indeed, it is only the massive limestones that are prone to karstification, and not chalky or thinly bedded limestones. The fractures channel the movement of water through the rocks, the permeability of which is slight. In these circumstances erosion is naturally concentrated on the sides of the fractures, and eventually underground rivers and caverns may form along them. This is why, in relatively juvenile karsts, in which solution has only as yet removed small quantities of the rock, the caves are often straight and angular, reflecting the fractures along which they have formed. The clarity of this adaptation naturally diminishes as the caves become larger.

Faults also have another influence on karstification, for they intercept the flow of underground water. Water that is following a bedding plane or a joint may come up against a fault which obstructs its path and ponds it back. If the fault is an open fracture the water may change direction and follow it. Faults thus interpose barriers to the movement of underground water, and divert it, and this happens particularly in fault-guided valleys and in karstic conditions. Every open fault in limestone tends to localise a

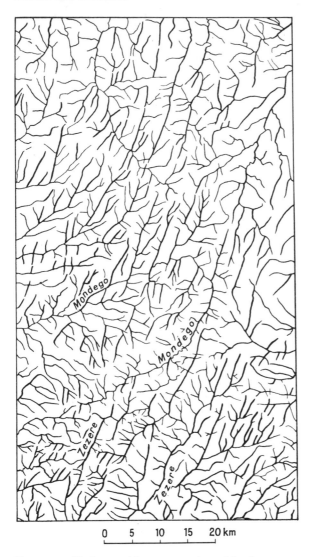

FIG. 4.23. Hydrographic pattern oriented by fractures,
in the granite massif of the Guarda region in Central
Portugal (*from Feio and Brito, 1950, p. 261*)
　　Stream-courses are aligned in two directions,
SSW–NNE and N 10° E. Many right-angled bends,
and many anomalies of texture in the stream pattern
(e.g. the interfluve between the Mondego and the
Zezera)

gulley or a valley or to produce some kind of anomaly in the stream pattern.
In the Paris Basin, for example, isolated incised meanders in the limestone
plateaus are often found where the river crosses a fault running more or less
at rightangles to its course.

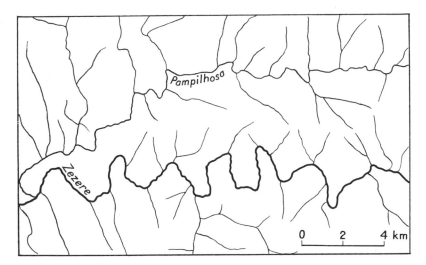

Fig. 4.24. Fault-guided meanders of the middle Zezera in central Portugal
(*from Feio and Brito, 1950, p. 259*)
  Such sinuosities are sometimes confused with entrenched meanders. In fact
they are closely guided by faults that produce their angularities. The influence
of the prolongation of these faults is to be found in the course of the
Pampilhosa, with its abrupt elbow-bends—but again these are different in
form from entrenched meanders

Naturally enough, this effect of faults on underground water is most
marked when the strata have been stretched by the faulting. In the case of
shatter belts, however, the accompanying tectonic breccias may often be
more resistant than the original rocks. Even if there is no resistant breccia,
the walls of the faults will be pressed together and will thus hinder underground circulation.

Induration by faults

Faulting may produce local hardening of the rocks in two ways:

1. When there are compact and well-cemented fault breccias. This occurs
usually in the case of compressional faults. The rock is crushed into fragments by the movement of the two sides of the fault plane, as underneath
overthrusts and very oblique faults. At some depth, recrystallisation may
occur, especially in limestones, and the crushed material is cemented into a
breccia. Such *fault breccias* are one variety of tectonic breccia. When the
cementation is very pronounced, so that no spaces remain between the
fragments, and the cement itself is crystalline, the fault breccia may be
much harder than the original rock, for it has neither stratification planes
nor fissures. It may thus be picked out by erosion in the same way as a dyke.
2. When magmatic products rise into the fractured zone. These may take
the form of thermal waters rich in dissolved materials that are precipitated

237

within the fracture, forming a hydrothermal dyke, or actual volcanic up-wellings. In both cases the open fissures of the shatter belt allow the rise of deepseated matter that fills the cracks and modifies to a greater or less extent the adjacent rocks, particularly if it is in the form of lava. These vein materials may be produced at varying depths. Basalt dykes, for example, may be formed quite near the surface, but the quartz veins that are common in crystalline rocks are of much deeper origin. Deepseated dykes in granitoid rocks may also be formed of crystalline material that differs in its crystal size from that of the rock into which it is intruded. In the case of pegmatites, the grain size is larger, in the case of aplites, smaller.

Local hardenings along faults, shatter belts and dykes are often brought into relief by denudation. To have much effect, however, they must be of sufficient dimensions, at least 100 m or more in width. But the width is usually irregular, and this is picked out in the relief, for when the hardened zone is thicker and stronger the resultant feature will be wider and higher, and when it thins out the relief feature does so likewise. Since the hardened zones are usually vertically disposed, the resultant ridges—whether result-ing from crush breccias or dykes—have ruined crests looking like the remains of a medieval castle. This effect is most noticeable with volcanic dykes. The Corions basalts provide a magnificent example at Roquemaure in the Rhône valley. In northeastern Brazil, there are some very fine dykes of pre-Cambrian quartz, that have been eroded into inselbergs in a semi-arid climate. Similar features are produced by aplite dykes that are of finer grain and less fissured than the granites that surround them. In this way faults of very great age, the ordinary effects of which have long since been eroded away, may still have an indirect influence on relief.

DERIVED SLOPES RESULTING FROM DIFFERENTIAL EROSION

In areas with rocks of contrasting lithology, faults may play a geomor-phological role for a long time after they have ceased to function. For example, a fault that brings into juxtaposition limestones and clays may contribute to differential erosion even after the original fault feature has been completely worn away. It only needs a renewal of erosion for this to take place, for streams will incise themselves more easily into the clay than into the limestone, and the valley slopes will develop more quickly. The erosion of the clay being more rapid than that of the limestone, the former forms a depression whilst the latter remains as a plateau. This is a classic case of differential erosion, just like that which ends up in the formation of cuestas. The zone of contact of the limestone and clay, along the faultline, is etched out, and a slope appears which is not of tectonic origin but is due to differential erosion. Such cases are called derived fault scarps or more commonly faultline scarps. The outcrop of the fault plane at the surface is known to geologists as the faultline, and a faultline scarp is one that is based on this faultline. It is a tectostatic relief form controlled by differences of lithology that produce differential erosion. This is why an escarpment so formed does not necessarily correspond to the tectonic displacement imposed by the fault on the strata.

In the case of horizontal or gently dipping strata, in which the effects show with the greatest clarity, the following types may be distinguished.

Reversed or opposed faultline scarps

In these the slope is in the opposite direction to that of the fault. There is an inversion of relief, for the upthrown side forms a depression whilst the downthrown side now stands out as in relief. Geographers have commonly called such features reversed faultline scarps. Unfortunately, however, the geologists give the name reversed fault to one which throws down the strata in the opposite direction to the dip, and so there is here a source of confusion which might be avoided by the use of the term opposed fault scarp (i.e. opposed to the structure).

Such opposed faultscarps are always due to lithological control of denudation and in the nature of things they are faultline scarps. They are particularly well developed in platform structures characterised by alternations of hard and soft beds, of sandstones and clays for example, or limestones and marls. If a fault scarp (as in Fig. 4.25) cuts through a bed of limestone underlain by marl, the downthrown side forms a limestone platform. The original fault scarp is a marl slope surmounted by a limestone cliff; it thus recedes easily, for most of the slope is composed of soft strata. If the dip is at an approximate angle the scarp will recede just like an ordinary cuesta, and a clay vale will develop at its foot. But if the whole area is sufficiently elevated above the regional base level, this clay depression cannot so persist, perched up as it is, without being denuded. It is then that the faultline scarp appears on the opposite side of the fault, as a result of differential erosion which removes the clay bed at the foot of the original scarp but leaves untouched the limestone on the downthrown side. The one fault is thus responsible for two scarps that face each other: the original fault scarp that conforms to the structure and the opposing one that is a faultline scarp. Its erosion is in no way connected with the throw of the fault, and depends only on the process of differential erosion, on the thickness of the soft stratum and the downcutting of the streams.

Opposed fault scarps may thus appear, provided that the lithological and morphogenetic conditions are suitable, when the original fault scarp has receded sufficiently. This implies, however, that the fault is no longer active; otherwise they would remain dependent on the faultline. They can thus only form during the period when the residual fault scarp is being dissected. An excellent example is to be found in the Rochepot fault in the Montagne Bourguignonne (see geological map, *Beaune*, scale 1:80 000).

The production of an opposed fault scarp does not necessitate the previous levelling of the faulted area by a widespread erosion surface. It is

239

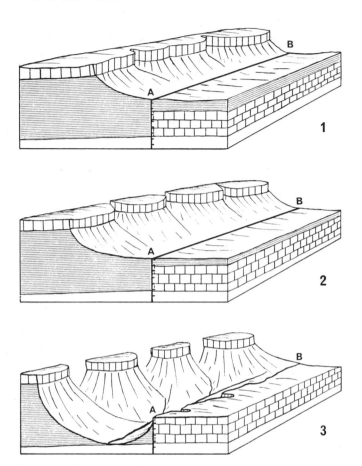

Fig. 4.25. Complex evolution of a fault-scarp

　1. The fault AB gives an original fault-scarp. Small streams begin to scallop the front of the uplifted block, and screes cover the marls beneath, concealing the actual fault. This would happen in a semi-arid climate.

　2. The same evolution continues, with no important change of base-level. The limestone capping of the uplifted block is more and more cut up by streams dissecting the original scarp, whilst the marl cover of the downthrown block becomes thinner underneath the lower slopes.

　3. A change of base-level, which may be due to a simultaneous uplift of both blocks without a renewal of faulting, produces a marked accentuation of the dissection. The limestones of the downthrown block are stripped of their marl cover. A longitudinal valley, following the line of the fault, erodes the marl outcrop on the edge of the uplifted block and partly exposes the fault itself. A second fault-scarp thus appears, but inversely to the structure, and this is termed an opposed fault-line scarp

enough that the original scarp should have receded sufficiently for the soft rocks at its base to be vigorously eroded. The retreat of parallel slopes in dry areas is, it would appear, favourable to such an evolution. It seems that this was the case in the Pliocene, with the original Rochepot scarp. The ultimate formation of the opposing scarp resulted from the combined action of the continued uplift of the Montagne Bouguignonne and the increasing headward erosion of streams from the plain of the Saône.

It is, however, clear that the truncation of a faulted area by a wide erosion surface, a pediplain for example, offers favourable circumstances for the production of faultline scarps when the surface is subjected to renewed erosion.

Subdued faultline scarps

When a fault has been planed off, and renewed denudation has initiated differential erosion, the faultline scarp that thus develops may coincide with the original appearance of the fault scarp, without any inversion. This is a normal or conformable faultline scarp. For such a situation to occur, the beds outcropping after the planation on the upthrown side must be more resistant than those on the downthrown side. The dissection resulting from renewed erosion thus scoops out the downthrown side and leaves the upthrown side upstanding. The new faultline scarp is thus an exhumation, as it were, of the original fault scarp. But the development of this new scarp is insufficient to equal or exceed the throw of the fault. Some part of the thickness of the soft beds that encouraged the exhumation of the scarp has generally been removed during the planation that is a necessary pre-requisite of this kind of landform evolution.

Exhumed fault scarps, which are but a variety of the faultline scarp, are usually subdued features of much smaller dimensions than the original throw of the fault. They are often found on the edges of graben which have been filled with strata of small resistance like sands, gravels, lacustrine alluvia and clays—in the Limagne, for example, at the junction of the crystalline rocks and the Oligocene marls. They are common, too, on the edge of horsts, for along the bordering faults there are often the remains of a cover of less resistant rocks, on the downthrown side, as in the case of the Liassic clays of the Bazois, west of the Morvan granite mass.

Exaggerated faultline scarps

Opposed faultline scarps may also be magnified, if the denudation exposes them to a greater depth than the original throw of the fault. Exhumed fault scarps cannot be magnified; they can only develop to the extent of the throw of the fault. Indeed, in order that they may develop at all the original fault scarp must be planed off to a level at which the soft rocks are on the downthrown side and the hard rocks on the upthrown side of the fault. On the other hand, an opposed faultline scarp may exhibit a degree of denudation much greater than the original throw of the fault, which in any case is reversed, since the upthrown side is opposed to the escarpment. All that is necessary is for the fault to have a throw slightly greater than the thickness of a thin hard bed overlying a thick soft bed, the thickness of which is greater than the throw of the fault. If the denudation is sufficient, this soft bed may be almost completely worn away. Thus the scarp slope has an altitudinal dimension greater than the throw of the fault. It is clear that exaggerated fault scarps are always opposed faultline scarps.

One important factor controls the evolution of all faultline scarps, and that is the relation between the throw of the fault and the thickness of the hard and soft strata. Its role is especially great when the rocks consist of alternations of hard and soft beds. A single fault, of varying throw, may sometimes give rise, after being planed off, to opposed scarps and sometimes to exhumed scarps of varying size. This is the case with the Mont Vigne fault, on the NW edge of the Morvan (see the geological map, *Avallon*, scale 1:80 000), which increases its throw towards the south. At this end, where the throw is greatest, the Jurassic limestones are brought up against the granite block. Although the fault has been planed off, renewed erosion has given rise to no scarp because the rocks on either side are of comparable resistance and there has been no differential erosion. On the other hand, the throw diminishes towards the north, and Liassic clays are

FIG. 4.26. Faults on the eastern margin of the Causses, near Trèves
The faults have been levelled by the surface of the Causses, and they appear simply as displacements in the lines of limestone crags

faulted against the granite. Here the fault has been exhumed, and a small north–south valley has been etched out in the clays on the downthrown side.

Faultline scarps present one other difference from original fault scarps. Since they are a product of differential erosion, they can only occur where the rocks offer sharp contrasts in resistance, as on the continental platforms. They are not found in shield areas, where the rocks are too homogeneous, nor in geosynclinal areas. In most cases they owe their origin to the renewed dissection of a more or less extensive peneplaned surface. The hydrographic regime is only partly adapted to the structure, and many streams are superimposed from the planated surface; it is these that, partially adapting themselves to the lithology, create the faultline scarps that appear during the course of the new cycle of erosion. The exposure of the slope of the fault plane is never perfect; thin layers of the soft formations will remain at the foot of the scarp as it develops, and the slope will never be as clearcut as that of the original fault scarp.

Faults can thus give rise to a very varied series of relief forms. The most striking are the consequence of the vertical movement of the beds on either side of the fault. Transverse faults, however, with horizontal displacement, may also influence relief, because of the shatter belts by which they are generally marked, which tend to form fracture-valleys. The role of faults is especially well marked in regions of more or less horizontal rocks, for in folded areas the faulting tends to become confused with the effect of the folds. Lastly, faults sometimes aid volcanic action—and it is to this that we next turn.

The geomorphological role of vulcanism

Volcanoes influence geomorphology in two different ways. In the first place, they create original landforms that result directly from volcanic action. These are mostly constructional forms, accumulations of volcanic products; but this is not always the case, for some eruptions also involve explosions, and these too produce characteristic landforms which, like the constructional forms, are attacked by erosional forces that gradually sculpture them. All these original landforms have one thing in common, the rapidity of their formation. For the most part they appear almost instantaneously—at least in terms of geological time. The interaction between internal and external forces is thus reduced to a minimum. The Davisian scheme, which may be false in other respects, is here satisfactorily applied, with two successive stages: the formation of the feature as a result of internal forces and its increasingly rapid modification by the forces of erosion.

Secondly, volcanoes exert an indirect influence, through the medium of lithology. Volcanic action causes the emplacement of new material which contributes to the earth's crust and persists for long ages after the vulcanicity has ceased. Ancient lava flows, dating from the Trias and the

Permian, and intercalated with clays, sandstones and conglomerates, influence the relief of the Boston lowland, as of the Saar. In southern Brazil, huge thicknesses of basaltic lavas of the Botucatu series contrast with the sandstones with which they are interbedded, giving clifflike scarps. When poured out at the surface, volcanic deposits are subject to erosion; but they are not always so disposed, and those that are, as it were, fossilised within sedimentary rocks may through their lithology give a characteristic flavour to the relief. And lastly, the magma from beneath may sometimes penetrate pre-existing sediments in the form of sills and laccoliths, the products of abortive volcanoes that never succeeded in reaching the surface. The ultimate relief forms are influenced by the lithological contrasts that such rocks provide. It is the same with volcanic plugs and dykes. Vulcanism survives then, on the geomorphological map, through the lithological influences that it exercises. These influences differ from those of both sedimentary and crystalline rocks, for the geometry of volcanic rocks is quite different.

We begin our study by an examination of the nature and origin of volcanic rocks, linking them to the character and evolution of the land forms that they produce.

Volcanic deposits and their formation

Volcanic rocks are one of the two varieties of magmatic products, the other being the plutonic or granitoid rocks. Petrographers have used the term *vulcanites* to describe the whole range of volcanic rocks, as opposed to the *plutonites*. Vulcanites in general are produced through eruptions that eject magma from a depth of several kilometres beneath the surface. This ejection involves a very abrupt change in the physico-chemical state of the material in which many important transformations occur. Volcanic rocks are thus of very varied character. These petrographic changes are accompanied by modifications in the physical properties of the materials, which in turn influence their deposition. Thus according to the temperature, lavas are more or less fluid, and may even give place to scoria. The mechanisms controlling the differentiation of magma and its deposition are thus much more varied and subject to change than those that operate in the case of plutonites, which are formed at great depth. The results are of great geomorphological importance, for they control the lithology of the vulcanites and the form taken by their accumulations.

MAGMATIC DIFFERENTIATION

Three groups of factors combine to differentiate magmas:

Structure of the earth's crust

We may recall the distinction outlined in Chapter 1 between sial and sima. Beneath the great oceanic basins, the lower layer of the crust is at shallow depth. In the continental areas it is overlain by material of a different nature and lower density. Basaltic magmas thus come from the lower part of the crust, whereas the plutonites are formed at the base of the upper layer. The great basaltic outpourings are thus typical of the ocean basins, granitisation and metamorphism of the continental areas.

On the basis of this concept, geophysicists have defined an 'andesite line' separating the regions with basaltic outflows from those in which the vulcanites are more acid. This notion is well applied to the Pacific and Indian oceans. The volcanic chains on the edges of continents, like the Andes, Kamchatka or the island arcs of eastern Asia, are characterised by outpourings of 'Pacific' type magma that is only exceptionally basaltic and is more commonly composed of dacite, rhyodacite, and very often rhyolite and andesite. On the other hand, the volcanic island arcs in the middle of the ocean basins have basaltic lavas, with trachyte and phonolite. Petrographers, somewhat unfortunately, have given the name 'Atlantic magma' to this suite.

We must not take the idea of the andesite line too literally; there are indeed very important basaltic outpourings on continental areas, which are of very much greater extent than those of the islands. As against Iceland, Hawaii and Kerguelen, the principal basaltic archipelagos, we must set the vast lava plateaus of the Columbia-Snake, in northwestern United States and southwestern Canada, of Ethiopia, the Deccan, and southern Brazil, which are of various ages from Triassic to Pliocene. The term 'Atlantic magma' is thus a peculiarly unhappy one, resulting from over-hasty generalisation. It would seem that the geophysicists have not taken adequate account of the structural factors, which have played a great part in the differentiation of magmas.

It appears indeed, that the two extreme types of vulcanites, basaltic and rhyodacitic, each correspond to a certain type of structure. The island arcs of the ocean basins and the volcanoes accompanying the rift valley systems of the plateau blocks are always of basaltic type. This is the case in central France, for example, in middle Germany, and in Scotland. The plateau lavas poured out from fissure eruptions are also basalts. True, there are other sorts of volcanic rocks, but they occupy but a small volume; they usually appear at the end of the volcanic activity, after the basalt flows. Such an occurrence is seen in central France, where basalts cover large areas in Aubrac, Velay and Coirons, in comparison with which the little volcanoes of Auvergne, formed later of different materials, are mere pimples. In the Rhine rift valley, where volcanic activity is of small significance, the Kaiserstuhl contains but little basalt. It seems, then, that we may assume that basaltic outpourings are formed by deep cracks in the sial or in the ocean beds which permit a rapid rise of a quantity of sima. Conversely, the vulcanites of dacitic and andesitic types are most abundant in the folded mountain chains of the geosynclinals that have been evolving for a long time—the Romanian Carpathians, the Andes and the rest of the circum-Pacific chains. Volcanic emissions, usually submarine, that represent the beginning of geosynclinal evolution, are normally of ultrabasic type.

The role of gas

Volcanic eruptions are always accompanied, to a greater or less extent, by the emission of gases. Sometimes, indeed, gas and steam constitute the only signs of activity, as in the case of *fumaroles*. The gases emanate from two sources. Some of them are of deepseated origin and are the result of chemical reactions within the magma. Others are simply water vapour derived from the water content of the rocks traversed or from underground watertables vaporised by the heat of the adjacent magma. Water that has filtered downwards to great depths, heated in this way and partly vaporised, is ejected under pressure and may reach the surface in the form of geysers. Often, it is true, these waters, very hot and mixed with volcanic gases, may be corrosive and may contain abundant dissolved material of great variety. A whole series of characteristic reactions is thus produced in the rocks with which they have contact, known collectively as hydrothermal metamorphism; these are different from ordinary chemical weathering and are localised along the fissures through which the hot water circulates. In limestone, the phenomenon known as hydrothermal karstification is produced. On reaching the surface, the hot waters precipitate some of their dissolved contents by reason of cooling and the release of pressure. This material is known as hydrothermal tufa and occurs, for example, around the vents of geysers.

Emissions of volcanic gases depend on several factors, chemical reactions within the magma, vaporisation of underground water, and the release of gases derived from the magma. In fluid magmas, bubbles of gas form and escape more readily than in the case of the more viscous magmas, in which the gas bubbles remain imprisoned, forming inclusions in the solid rock.

These gas emissions are important for two reasons. In the first place they modify the chemical composition of the magma and thus influence the nature of the rocks that eventually consolidate therefrom. Secondly, they have important mechanical consequences. An emission of gas during the process of cooling hinders crystallisation and results in the production of irregular and ill-formed masses of scoria and cinders. Gases concentrated in certain parts of a volcano and submitted to increasing pressure may provoke explosions that overcome all obstacles; this explosive activity is due to the obstruction of the gas outlets by solid or viscous materials.

These explosive phenomena, which are of such geomorphological importance, are controlled in part by the nature of the chemical reactions produced in the magma during its ascent from the depths, and partly by the resistance the magma offers to the escaping gases. Both these factors depend on the chemical composition of the magma, because this is what, at a given temperature, controls the fluidity and so the resistance to gas pressure.

Basaltic magmas, which are more fluid, generally allow gas to escape without much difficulty; except when they are rendered pasty by cooling, in which case scoria may form, giving rise to explosions. But such explosive activity is much more important in eruptions of acid magmas, and so it is especially significant in the later stages of vulcanism; and acid magmas are often scoriaceous and cindery.

The role of temperature

Temperature also plays an important role, in combination with the chemical composition of magmas. In general the hotter the magma, the more fluid it will be, and thus more easily flowing and parting more freely with its gaseous content. The outpouring of lava will proceed smoothly. On the contrary, a sticky magma hinders the escape of gas and so the eruption assumes a more explosive character.

For every magma there is a critical temperature at which crystallisation will start, and when this is reached in cooling, the magma solidifies. As the effective temperature of the magma approaches this consolidation limit, it becomes more viscous and less easily flowing. The critical temperature is a function of the chemical composition; it is higher for acid magmas that are rich in quartz, lower in basic magmas. It seems also that the slight viscosity of basic magmas is due essentially to their high iron content, which renders them but feebly refractory. At the temperatures normally reached in eruptions, however, acid magmas have slow molecular diffusion, and this hinders crystallisation and keeps them in a pasty condition with a high viscosity. Since crystallisation does not take place at a high enough temperature, consolidation happens suddenly at a lower temperature under surfusion conditions, giving a glassy structure. If the material is pulverised by the explosion of the imprisoned gases, it forms scoria and cinders.

The low viscosity of basic magmas allows them to remain fluid and to flow easily at a wider range of temperatures and especially at lower temperatures. And at the same time they are more easily stirred up, so that the escape of the contained gases is easier. It seems, however, that many of the reactions that facilitate the escape of gas are themselves productive of heat, i.e. they are exothermic. The very fluidity of the lava thus helps it to maintain a higher temperature, or to put it another way, helps to slow down the rate of cooling and so to maintain the fluidity. Thus the fluidity is prolonged by auto-excitation; and this makes it easier to understand how basaltic lava flows have spread out over such vast areas. Such a system of auto-excitation comes into play especially, it seems, in magmas that are rich in iron and so naturally more fluid. But it also occurs sometimes in other magmas, as in the case of the extraordinarily regular beds of andesite at Infernillo in the Andes of central Peru, which from a distance look like marine limestones, so even is their stratification. The same phenomenon can perhaps offer an explanation for the bedding of certain sills, which have been extensively intruded into sedimentary rocks, and have succeeded in inserting themselves between the strata with no more disturbance than the mere separation and stretching of the sedimentary layers.

Of all the factors that contribute to the differentiation of magmas, the most important, it would seem, are those that, given the conditions of access to the surface, control the rate of cooling and the disposal of the imprisoned gases, and in consequence the maintenance of heat through exothermic reactions. We can understand, therefore, why vulcanicity is influenced by broad structural circumstances and why it is different in geosynclinal mountains, in the rift valley systems of continental blocks, and on the ocean beds. But although of great importance, this is not the only factor. During the course of development of volcanic action in any given region, the conditions under which magma reaches the surface will be modified. After a while, consolidated magma will block up some of the vents; its upward surge will also diminish, and it will encounter more and more obstacles that it is unable to overcome—and the eruptions will eventually cease. But this will not take place suddenly: the magma, hindered in its upward passage, will solidify more quickly, and becoming less fluid it is more liable to explosive activity. This is why such manifestations are frequent towards the end of a volcanic episode, even in regions with basic magmas.

PRINCIPAL TYPES OF VOLCANIC DEPOSITS

The products of vulcanicity can be classified from two different points of view: their chemical composition, and their physical condition. Although the chemical character of rocks has some effect on the weathering processes, it is their physical nature which is on the whole more important geomorphologically. However, the two aspects are themselves connected, for acid lavas, poor in ferromagnesian content, are more viscous at a given temperature than basic lavas and often solidify in a different way.

Petrographical classification

Two factors enter into this mode of classification: texture, that is the crystalline form, and chemical composition.

Dealing first with texture, we can distinguish, amongst the volcanic rocks, those that consolidated in two phases, so that they have a microlithic texture, with both large and small or microscopic crystals embedded in a glassy matrix—like rhyolites, basalts and phonolites—and the glassy rocks that crystallised quickly in a single phase so that crystallisation could go no further than to produce a fine matrix with a glassy or conchoidal fracture (provided that gas bubbles are not too numerous). Granular or holo-crystalline rocks, formed entirely of crystals, consolidated and crystallised slowly and at great depth, are of plutonic origin, like granite and syenite. But although controlled by the circumstances of consolidation, texture is not directly related to mode of origin. Lava flows may yield glassy as well as microlithic rocks; but the largest flows are of microlithic rocks like basalt. Explosive action may result in microlithic rhyolites, for example, as well as in glassy rocks, cinders and pumice. It is important, therefore, for the geomorphologist to understand the classification in order to be able to identify rocks petrographically. However, texture also influences weathering. Both microlithic rocks and granites are subject to granular disintegration, as we find, for example, under periglacial conditions at high altitudes in Peru, with both the ignimbrites of Huaron and the rhyolitic tuffs of the Huenque River. It is true that the texture of these rocks is microgranular rather than microlithic. With typical microlithic rocks like basalt, granular disintegration cannot take place because of the virtual absence of phenocrysts—large crystals, visible to the naked eye, of olivine, for example. This is why, in humid tropical climates, basalt does not decay in the same way as granite; it weathers directly into clay, but relatively slowly and not very deeply.

Secondly, from the point of view of mineralogy, we can classify volcanic rocks according to their acidity, that is their silica content. When silica is abundant it forms quartz crystals, made only of silica. If it is less abundant, it only appears in combination with other elements, giving silicates, of many varieties and different crystalline forms or simply forming the glassy matrix. Then there are the rocks that contain no silica at all, not even in combined form. The silica content only has an indirect influence on the fluidity of lava. A silica-rich rock is necessarily poor in ferromagnesian minerals and vice versa. If the ferromagnesian content is high, then silica is low and there is no free quartz. But it is not the vulcanites that are rich in ferromagnesian minerals that are the most basic, far from it, and this is why acidity is not the most appropriate criterion to adopt for the understanding of the mode of formation. It is only useful in the case of the most acid vulcanites, for these are less fluid than rocks rich in ferromagnesian minerals, which are less acid. But rocks rich in ferromagnesian minerals are more fluid than the very basic rocks which have no felspar and no felspathoids.

We can thus classify vulcanites on the basis of their decreasing acidity, as follows:

1. The most acid rocks, with quartz crystals, are the *rhyolites*, in which potash felspars are dominant, and the *dacites*, with plagioclase felspars. Of the glassy rocks, *obsidian* contains free quartz.

2. Rocks with a moderate amount or a deficiency of silica, in which there is no free quartz, but silica in the form of felspathoids, include *trachyte*, with potash felspars dominant, *andesite* with acid plagioclases, *labradorite* and *basalt* with basic plagioclases. Labradorite and basalt are the most fluid of all the volcanic rocks. *Phonolite* comes next from the point of view of degree of acidity.

3. Lastly, the ultrabasic rocks do not even contain felspars—such as

nephelinites, limburgite and kimberlite, with which diamonds are some-times associated.

These vulcanites correspond to different types of holocrystalline or microcrystalline plutonic rocks: rhyolites are the equivalent of granite, dacites of quartz-diorite, trachytes of syenite, andesites of diorite, and labradorites and basalts of gabbro.

Classification by physical form

This is much more important, geomorphologically, for it directly controls the behaviour of the rocks in relation to weathering agencies. The physical form of the rocks depends on the circumstances of their emplacement, and we can distinguish several broad categories.

Lava flows are outpourings of lava that is sufficiently fluid to flow. The flow is usually laminar, for the viscosity is too great to allow of much turbulence. In this way the superficial layer remains at the top, and this allows it time to cool, after the escape of the gases, which, whilst they are actually burning, may give the atmospheric layer immediately overlying the lava a temperature higher than that of the flow itself. It thus forms a cindery crust or congealed skin on the flowing lava. The flowing move-ment causes frequent rucks or fractures in the crust, and these may remain as microrelief features after the flow has ceased. These chaotic lava sur-faces are given the name of *cheire* in the Auvergne, and are called *aa* in Hawaii. American authors use the latter term for the disjointed slabs of consolidated crust; folded and contorted surfaces are called ropy lavas, or *pahoehoe*.

Each outpouring, corresponding to a particular phase of the whole eruption, gives an individual lava stream or *coulée*. It is generally sharply distinguished by its stratification: its base rests on the crust at the top of the preceding flow, or maybe on the soil that formed thereon in between one eruptive period and the next, and which is often baked to a red colour. In very fluid lavas, the consolidation may permit the formation of hexagonal columnar prisms, particularly well developed in basalts. This columnar structure, perpendicular to the plane of cooling, aids the subse-quent weathering of the rocks.

In the case of the fluid lavas, the edges of the flow, which are in contact with the solid rock over which the lava stream is moving, become solidified before the middle, which remains liquid and continues to flow. The whole *coulée* may thus become almost literally disembowelled, and this leaves behind long sausage-shaped hollows, elongated in the direction of the flow. Such tubular features may be brought to light later on by the collapse of their roofs, or by the general dissection of the lava stream; they have been described in the Middle Atlas of Morocco, and in Victoria (Aus-tralia).

Each lava stream thus constitutes a bed, more or less massive and perhaps with columnar structure, according to the circumstances under which it cooled and consolidated. Since lava can only flow if it is in a sufficiently thick stream, each *coulée* may be several metres thick, perhaps as much as 10 m; and such lava beds will play an important geomorphological role by reason of their hard massive and less fractured character in comparison with equivalent thicknesses of sedimentary rocks. They are particularly liable to form crags and rock ledges.

Clastic materials, principally cinders and ashes, are expelled from volcanoes by more violent action of a more or less explosive kind. Pumice is also frequently formed; it consists of glassy material very rich in included gas, which renders it very light. Cinders may be either microlithic rocks or glassy rocks, shattered by explosion.

The essential geomorphological characteristic of this class of volcanic products is that they form mobile deposits. Cinders, ashes and pumice behave more or less like unconsolidated rocks of similar particle size, with some difference however, especially in the case of pumice by reason of their very low density. For example, they often float on water, which makes their transport by rivers very easy. In alluvial accumulations, because of this property, they tend to get deposited as banks on top of more normal sandy material.

The particle size of these mobile volcanic products varies from the fine grain of a silt formed by fine ashes to a coarse conglomerate of bombs. But the most abundant materials resemble silts and fine sands. The finer particles may float in the air for a long time in the form of dust, as happened for many months after the explosion of Krakatao in 1883, for example. They then contribute to the formation of the aeolian sediments known as *aeolocinerites* (aeolian tuffs). Thick accumulations of this kind may have important geomorphological consequences, as we shall see later on. Cinders, however, are easily reworked by running water into *hydrocinerites* or by mud flows into *ludocinerites*.

All cinder eruptions, however, do not yield clastic materials. Explosions of Peléan type, that give rise to *nuées ardentes* or burning gas clouds, throw out very hot, half-molten cinders, that have no time to cool in the atmosphere before being deposited, since they are borne in a cloud of incandescent gas. The soft and partly molten cinders coagulate and consolidate into *ignimbrites*, which look rather like a sandstone with crystals and grains set in a glassy cement. These ignimbrites are subject to granular disintegration. They generally form thick and massive beds which only break with difficulty into large blocks; their surface is often characterised by a jumble of tors and piles of blocks that look like ruins.

Eruptions of material that has but slight fluidity often yield scoriaceous products, irregularly consolidated, looking like a curdled mass at a distance but giving a very unevenly aggregated appearance at close quarters. Massive chunks alternate with more scoriaceous material, and the latter is often poorly consolidated as a result of being deposited in a pasty condition.

251

Such materials are poorly and above all irregularly resistant to erosional forces. The least consolidated parts are most easily attacked, whilst the more coherent masses remain as upstanding stumps. When they alternate with true lava flows, such volcanic products behave like soft beds, allowing the overlying harder lavas to overhang in cornices.

Lastly, we may note that cinerites, and especially the aeolocinerites and ludocinerites, consolidate very easily after deposition, producing the friable rock known as *tuff*.

Volcanic products thus present, according to their physical state, a whole range of possible reactions to morphogenetic processes. They may be rocks of great resistance, like certain lava flows, plugs and sills, or unconsolidated and powdery materials, easily removed by the wind. The lithological factor is thus of prime importance when vulcanites are worked upon by the forces of erosion.

MODES OF DEPOSITION OF VOLCANIC MATERIALS

Following the famous French petrologist Alfred Lacroix (1883–1948), we may essay a classification of volcanic eruptions, and then examine the actual activity of volcanoes, which often consists of a series of eruptions of different types.

The Lacroix classification of volcanic eruptions

There are four types as follows:

The *Hawaiian* type is characterised by very fluid lavas, poured out more or less continuously. The crater is occupied by a lake of molten lava, and the gases that escape from this reservoir burn continuously, thus keeping the surface temperature at a higher level than that of the molten mass below (in 1917, a temperature of 1 350°C was measured in the flames as against 750–850° in the cindery skin on the surface of the lake). From time to time the lava boils up under the influence of the escaping gas and over-flows, resulting in a *coulee* or lava stream that flows down the side of the cone. This type, then, is characterised by the continuous emission of very fluid lava in small quantities.

The *Strombolian* type (from Stromboli, a volcanic island to the north of Sicily) is also characterised by permanent activity, but the material is less fluid. There is no lake of molten material, no lava flows. The eruptions are of cinders and stones made of lava-covered cinders, which are ejected in a series of small gas explosions. In Central America, in Salvador, the volcano Izalco was of this type, until its activity ceased in 1958. The ejectamenta form a dust cloud and then fall back on to the margins of the crater, thus building up a conical mass of pyroclastic material.

The *Vulcanian* type (from Vulcano, near Stromboli) is characterised by explosive activity of a more intermittent type. The vent becomes choked with solidified lava which impedes further emissions. In between the paroxysms the gases escape through lateral fissures. When a paroxysm occurs the plug obstructing the vent blows up and pumice is ejected, forming a mushroom cloud of dust. Small lava flows may occur, but not being very fluid, they do not get very far. Such lavas are in part rhyolitic.

The *Peléan* type, called after Mont Pelée in Martinique, is much rarer than the others. It is characterised by extremely violent explosions, very high temperatures, and lava that has very little fluidity. A tall plug of coherent but very viscous rock forms in the crater and rises gradually through gas pressure at its base; its sides are fluted as a result of friction. The gas pressure that causes it to rise renders the whole situation unstable, and there comes a moment when the gases find a lateral outlet by violent explosive action. It was in this fashion in 1902, that a stream of incandescent gas, laden with volcanic dust at a very high temperature, higher even than that of the flames above the lava in the Hawaiian crater of Halemaumau, swept like a wave, destroying everything in its path, upon the town of St Pierre. When the debris is deposited, it aggregates into ignimbrites. Such eruptions, of great violence and destructiveness, though happily of some rarity, are known as *nuées ardentes* or burning clouds.

It must be emphasised that Lacroix only classified types of eruption, and that in general, the larger volcanoes and more particularly the volcanic complexes, experience successive eruptions of different types. The Lacroix classification is useful only if we bear this in mind. Outpourings of very fluid lava may obscure the fissures from which they emanate; they do not come from sharply defined craters but from numerous cracks that traverse an area, without the formation of individual cones. This is the case with the basalts of Velay and the Coirons, in France. Vulcanian explosions are often more sudden and more important, resulting in the partial destruction of the volcano, which is shattered, sometimes without ever having had a separate existence. They form explosion craters, either at the expense of a pre-existing cone, as in Vesuvius, or without the presence of a cone, as in the *maare* of the Eifel. However, these diverse types of paroxysm come within the Lacroix classification; they only differ from the examples chosen by Lacroix in their violence and intensity—and this leads us to an examination of the rhythm of vulcanicity.

The rhythm of vulcanicity

At this point it becomes necessary to round off the Lacroix classification, for this is based solely on the nature of the volcanic phenomena without regard for either their intensity or their periodicity. It groups highly catastrophic episodes, which may be unique in the life of a volcano, like a peléan-type eruption, with the more permanent phenomena characteristic of the strombolian and hawaiian types. But even volcanoes of these last two types also have paroxysms. From time to time in Hawaii, for example, great lava streams form, flowing right down to sea level and covering millions of hectares. In this case the Lacroix classification represents the permanent or background type, which is occasionally interrupted by more violent episodes of the same general character.

From a geomorphological viewpoint the rhythm of volcanic action is very important, for the various phenomena produce landforms, either constructional or resulting from explosions. Between the outbursts, erosional forces proceed to remove their effects. Cinder cones like Izalco are very easily eroded, and if they cease to be constantly added to they rapidly disintegrate through the ravinement of their slopes. From the moment that volcanic activity ceases to be continuous, therefore, there is a more marked interaction between the creation of new landforms by the volcano and their modification by external agencies of erosion. Ravines eroded in old lava flows may provide a passageway for a new flow. This interaction, which is so important for the understanding of volcanic relief, depends on the rhythm of the volcanic activity. Its role is the more important if major eruptions are separated by long periods of quiescence or minor activity.

In the same way, a succession of eruptions of different type from the same volcano may influence its relief. This is so in the case of Vesuvius, which has been intensively studied. The obstruction of the summit crater causes the appearance of lateral outlets along fissures in the flanks of the cone, which may give rise to subsidiary ash cones. A similar happening characterises Mauna Loa in Hawaii, but in this case the outlets produce basaltic lava flows. Vesuvius has a major crater, surrounded by a rim, forming Monte Somma, that was partly removed by the explosion which destroyed Pompeii in A.D. 79. Within this old crater a new cone has formed. Even during the course of a single eruptive episode, several different products may be emitted in succession. Lava flows, which often emanate from the flanks of the cone, are often accompanied by the ejection of ash, cinders and pumice by explosions in the crater itself. One could thus graph the eruption by plotting the types of product in terms of time along one axis and their relative abundance along the other.

An excellent example of the successive action phases within a single volcano is provided by the fissure eruption of Threngslaborgir, in Iceland, studied by Rittmann. At the beginning, the enlargement of the fissure that permitted the later eruptions was accompanied by earthquakes and the ejection of debris, and the edges of the fissure were broken. This allowed the emission of large quantities of very fluid basaltic lava which spread out in thin, flat layers. Ash eruptions followed, causing the formation of an aligned series of cones. During a fourth phase, pools of lava appeared in the craters, and these ultimately overflowed to give a fresh series of lava streams which in part covered the ash beds and the previous lavas. Finally, the activity died down and nothing is left but fumaroles.

Volcanic landforms and their evolution

The landforms resulting from volcanic manifestations are as varied as the eruptions themselves. Most of them are constructional forms, caused by the accumulation of vulcanites, but there are also what one might call negative forms, caused by explosive action. They also vary considerably in size and complexity. The simplest forms arise from one single eruption—lava flows, plugs, explosion craters for example. The complex forms on the other hand are created by a variety of forms of different ages.

SIMPLE FORMS

These are found in great variety according to the type of eruption and the character of the materials ejected.

Lava flows

These are streams of fluid lava that move with a laminary motion. Their speed decreases, in general, with distance from the source, owing to cooling. It may reach 30 km/hr in the early stages. Their width varies from a few metres to a kilometre or so, exceptionally more. Their thickness is always at least several metres, for a thinner lava stream would quickly come to a stop through rapid cooling. It is the lavas rich in ferromagnesian constituents, and in particular basalt, that are most apt to produce good *coulées* by reason of their fluidity, and in some regions vast areas hundreds or perhaps thousands of square kilometres in extent are covered by successive basalt flows, without any cones, the lava having been emitted in an extremely fluid condition from fissures. A good example is the Coirons area of France.

The lava stream often has a cindery layer at its base. As the stream moves forward, clinker forms on its front, and this gets overwhelmed by the moving lava behind it.

The surface of lava flows exhibits several forms of microrelief features; ropy forms (*pahoehoe*) when the pasty skin has been wrinkled by the movement of the liquid lava beneath, and *aa* made of blocks of clinker broken by the movement of the lava underneath and piled up against each other. *Aa* often appear when a narrowing of the lava stream accelerates its movement. *Pahoehoe* on the other hand, occur where the lava stream spreads out and moves more slowly. Localised gas emissions create blisters sometimes several metres high, forming cone-shaped masses of clinker resulting from an explosion. They often occur where the lava stream has crossed a damp patch which provided a quantity of water vapour that the heat of the lava converted into steam. Sometimes the explosion chamber is partly preserved as a cave.

The general shape and size of the lava flow are conditioned by the type of lava and the pre-existing relief. Viscous lavas only produce short, thick flows on steep slopes. Often they do not even reach a valley that might have channelled them, and they stand out like burrs on the flanks of the volcano, near to the point of emission. On the contrary, fluid lavas reach gulleys and valleys in which they flow, and sometimes, if the relief is not too dissected and if the outpouring is sufficiently abundant, they may succeed in 'fossilising' the underlying land surface completely. This is what has happened underneath the plateau basalts. When valleys are dammed up by lava flows, lakes may form. The cutting of gaps in these massive lavas by running water is a very long process, so that normal headward erosion is prevented for perhaps a considerable time. Alluvia may be spread over the lava surface, sometimes to be covered by further lava flows and so left stratified within the volcanic formations. Within a valley, a lava stream usually has a rounded cross-profile, rather like that of a glacier, and running water is thus concentrated in marginal gutters, between the lava and the valleyside; these marginal streams may actually succeed in incising themselves into the edge of the old valley which is fossilised underneath the lava. Such incision usually takes place much more readily into the country rock than into the tough basaltic lava.

Fig. 4.27. The peléan 'aiguille' of Petit Piton, in St Lucia, West Indies

Aiguilles and domes

At the opposite end of the eruption scale are the masses of highly viscous and practically solid lava that may be extruded by gas pressure. A peléan explosion is a sort of misfire of the volcanic mechanism. It is not strong enough and it is too slow, so that the mass of almost solid lava in the throat of the volcano is merely pushed upwards as through a die, thus producing either an *aiguille* or a dome, according to the consistency of the lava. The term *aiguille* is reserved for an almost circular extrusion with very steep, indeed precipitous sides. Such plugs, often crowned by churches, are found in Le Puy-en-Velay. The lava at the moment of extrusion was almost solid. The sides of the plug often exhibit flutings that look like fault planes and are evidence of the friction engendered by the extrusion. *Domes* have gentler slopes, of 30° or so. They result from the extrusion of a magma that was slightly more fluid and so had a greater tendency to spread out. Naturally therefore there are no friction flutings.

It is often difficult to distinguish domes that result from extrusion from the plugs or necks resulting from differential erosion, that we shall examine later on. The ground-plan of these extrusions is generally circular, like the volcanic chimney from which they emerged. But fingerlike projections may occur, sometimes aligned along fractures, thus complicating the plan. In St Lucia in the West Indies several *aiguilles* exist side by side.

Aiguilles, being the result of gas pressure, are often associated with scoria which, like burning gas clouds, are another manifestation of explosive vulcanism (as in Mt Pelée). For this reason it is sometimes difficult as in the case of domes, to distinguish between original forms and those that have resulted from differential erosion.

Pyroclastic deposits

These are the cinders and ashes resulting from volcanic explosions. The materials are shot into the atmosphere and fall back to earth at a greater or lesser distance from their point of origin, depending on their particle size and the force with which they were ejected. Clouds of volcanic ash may float around in the upper atmosphere for weeks, as after the explosion of Krakatao. Rains of ashes, blown by the wind, may cover thousands of square kilometres. In 1912 the eruption of the Katmai volcano in Alaska covered 7 500 sq km with ashes to a depth of 30 cm and 75 000 sq km to a depth of over 2·5 cm. At Kodiak, 170 km from the volcano, 13 cm of coarse ash fell on the first night after the eruption and 18 cm of fine ash during the next two days. Small ash falls were even observed at Puget Sound, 2 300 km away.

The ashes from each eruption have a definite composition and can usually be recognised by their mineral content. They are thus invaluable stratigraphical datum planes, identifiable in alluvia, colluvia, volcanic deposits, etc. Their use for this purpose is called *tephrochronology*.

257

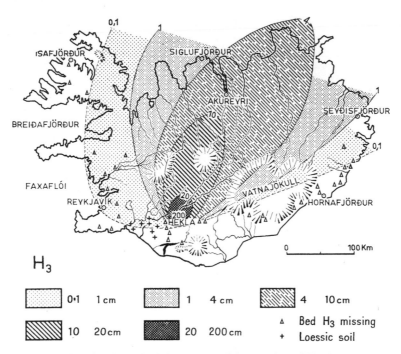

FIG. 4.28. Distribution of ash from a rhyolitic eruption of Hecla
(Bed H3) *(after Thorarinsson, Einarsson and Kjartansson, 1959, p. 154)*
    This eruption was one of the most important in post-glacial times.
The influence of south-westerly winds on the spread of the ash is
obvious

    Ash rains, accompanied by pumice and sometimes by rock debris consisting of broken rock and baked pebbles (if the eruption breaks through thick detrital deposits), create carpetlike surfaces of low relief that are very easily eroded. The ashes, light and fine-grained, are easily moved and carried away, especially in the absence of any vegetation to fix them. As an example we may cite the Arequipa area of Peru, where ash rains covered an area of acid mountains, the surface of which has now been almost completely scoured whilst the redeposited ash is found only along valleys. The streams flow with clear water, and the bottoms of the larger valleys are marked by small terraces of hydrocinerites. On the contrary, in low-lying regions or basins that are badly drained—maybe because of a volcanic barrage—cinerites accumulate and may form thick infillings. There is such an occurrence in the central depression of Salvador, which is a rift valley bordered by volcanoes. Ash beds, with or without pumice, sometimes aeolian, sometimes deposited in water, alternate with palaeosols that were formed during the periods between the eruptions.
    Hydrocinerites are of frequent occurrence in volcanic foothill zones, where they often spread over wide areas. They are frequently associated

Fɪɢ. 4.29. Hydrocinerites, spread and deposited by water, in the island of
Sel, in the Cape Verde group
   The cinerites were spread by diffuse stream action around the base of
volcanic cones. The beds are characteristically pellicular. The white layers
are Quaternary calcareous crusts

Fɪɢ. 4.30. Valley fossilised by hydrocinerites, south of Valle Pescadores in
Peru
   The valley has been preserved by cinerites washed down from the slopes

with pebble beds derived from scoria. Such deposits occur in all sorts of climate. Ash eruptions may kill off the vegetation; and they are usually accompanied by torrential rains. Conditions are thus very favourable for erosion, for the destruction of the vegetation creates, all of a sudden, a pseudodesert whilst the violence of the rain causes exceptional floods. Morphogenesis of an extraordinary intensity is thus set in motion. The slopes of fresh aeolocinerites are deeply gullied, sometimes down to the rock beneath; and the torrents cease to flow after the catastrophe.

Cinerites, moreover, finegrained and porous as they are, are often thixotropic, and they are thus an easy prey to the flows which themselves are occasioned by the torrential rains that accompany the eruption. Sometimes mud flows are generated by the breaching of lava-dammed lakes; such occurrences are well known in Java where they are called *lahar*. But this term is extended to apply to all mud flows of volcanic ash associated with eruptions. Great thicknesses of ludocinerites may be formed in this way, sometimes incorporating boulders caught up from the subjacent rocks. *Lahars* move very quickly, and often dam up valleys. One of them in 1953 on the slopes of Mt Spurr, in Alaska, created a lake 8 km long. Sometimes too they spread out over the low ground at the base of the volcano. Around the volcanic Acajutla, in Salvador, their muddy deposits alternate

FIG. 4.31. Gradual scouring by stream action of a rain of ashes, south of Arequipa in Peru

The crystalline rocks are highly dissected, in an arid climate, and are coated with a dark patina. The ashes are of pale colour; they have been almost entirely removed by stream action from the ridges and are re-deposited in the ravines and in the main valley where they form a small terrace of hydrocinerites

Fig. 4.32. Deposits of a lahar, near Vieux-Fort in St Lucia, West Indies
At the bottom, finely stratified layers indicate the action of water; above, an unstratified mass, with much fine-grained material and a concentration of stones and heavy boulders at the base, is a typical exposure of the results of a boiling mud-flow or lahar

with hydrocinerites. Elsewhere there is a gradation from mud flow to ordinary deposition in water.

Scoria, which are less mobile, create less unusual forms. They fall, together with volcanic bombs, which are chunks torn from the walls of the crater and wrought in a pasty condition, around the hole from which they were ejected. Sometimes, mixed with ash, they may build up cones, like Izalco in Salvador, or annular ridges around a large vent which emits less material, as in the case of the *maare*. The particles are ejected at a high temperature, and when they fall, still in a pasty condition, they become welded together and thus very resistant to erosion. Beds of scoria may remain for a long time as flat or undulating surfaces, or as cones around the craters.

Nuées ardentes or burning gas clouds represent a halfway stage between pyroclastic accumulations and lava flows. Burning at a very high temperature, they move very rapidly, channelled by valleys, which they fossilise beneath pyroclastic materials cemented in a glassy matrix, massive in structure and very resistant to erosion. The exhumation of the valleys down which they have flowed is almost impossible.

Mud volcanoes

Mud volcanoes are a peculiar manifestation of certain forms of vulcanism, but they are not always of volcanic origin. They comprise accumulations of

261

mud in the form of a flattened cone around a vent, the mud being provided by clay formations at a greater or lesser depth underneath. In Baluchistan such cones may vary from 6 to 100 m in height, with a basal diameter of 50 to 230 m. In a desert climate that allows the mud to dry out quickly their flanks may have slopes of 30 to 40°. The mud is often mixed with sand and stones that are ejected at the same time.

Mud volcanoes are generally associated with gas vents, and the gas bubbles up through the pool of liquid mud in the crater. Sometimes the craters are aligned, somewhat resembling fissure eruptions. The mud that comes to the surface is rendered liquid by admixture with water and is urged upwards by gas pressure. The gas is sometimes of volcanic origin, as at Pozzuoli, near Naples; but other mud volcanoes are 'paravolcanic' manifestations, found especially in regions that abound in hydrocarbons—like Baluchistan. In these cases it is petroleum or 'natural' gas that causes the mud eruption, the mud being rendered fluid by admixture with saline water. But earthquakes, which set thixotropic reactions in motion, are also responsible for mud eruptions. Lines of mud volcanoes mark the position of fractures, and especially of active faults, which bring pressure to bear on deepseated clay deposits.

Destructive forms: craters

Craters are depressions created by volcanic action on the site of a vent. Two distinct mechanisms contribute to their formation:

1. Explosive action that sweeps the chimney, so to speak, and expels material from its walls, thus enlarging the orifice. Volcanoes that produce pyroclastic materials are particularly prone to this action, which in its most extreme form may even destroy the cone by blowing it to pieces, as at Krakatao. In volcanoes of the Pelée type, the obstacle provided by the plug causes explosion craters to open up on the sides of the cone. On the other hand, in volcanoes that emit cinerites and fine ash, the explosive activity is directed upwards, affecting the lips of the crater. Many intermediate varieties exist, however, depending on the resistance of the volcanic mass and the localisation of the gas pressure that provokes the explosion. Frequently the explosion will carry away part of the rim of the crater, or perhaps the summit of the cone, as happened with Vesuvius in A.D. 79. In this way a breached crater is produced; and when the crater itself is considerably enlarged by the explosion we speak of an explosion crater.

2. Subsidence resulting from the emission of lava—producing subsidence craters, whether the action be sudden or slow and gradual. Such craters are found in all volcanoes that emit large quantities of fluid lava. During their upward progress these lavas have often distended the rocks through which they have passed, sometimes even interpenetrating them in the form of sills and laccoliths. On eruption, some of this pent-up lava escapes and the magma chamber is more or less completely emptied. The result is a

FIG. 4.33. The crater of Misti, in the Peruvian Andes
The photograph shows the summit of the cone; on its flank, in the foreground, is a deep gulley or 'barranco'. The cone is composed of successive layers of pyroclastic materials, dipping away at an angle of 45° and giving very steep structural slopes. The crater itself is quite shallow

FIG. 4.34. Section across a crater (Sveinar, Iceland) (*after Thorarinsson, Einarsson and Kjartansson, 1959, p. 161*)
The chimney occupies a fissure that traverses a varied succession of lava-flows. The small cone is built of pyroclastic materials and lava-flows, breached by an asymmetrical crater

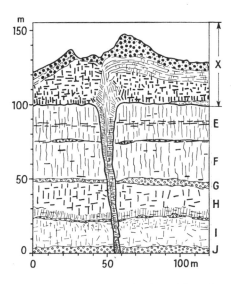

diminution in volume, which combines with the pressure induced by the weight of the surface accumulations to produce concentric fractures around the point of eruption, crossed by radial faults that separate blocks which may founder unevenly. Such features occur as the volcanic activity dies down, either temporarily or completely.

263

FIG. 4.35. The Soufrière volcano, Guadeloupe, in eruption in October 1956 (*drawn by Rimbert from a photograph by Lasserre*)
 The eruption is occurring along an easily-seen fissure, marked by the smoke emissions. This is an active volcanic dome, with the beginnings of radial dissection

Craters are thus closed depressions, surrounded by steep slopes that result either from subsidence or from the clearing action of explosions, or even, in the case of the most violent explosions, from the removal of whole chunks of the volcano. On the crater sides can be seen the stratification produced by successive layers of erupted material. In volcanoes with fluid lavas, the bottom of active craters may be occupied by a pool of lava. When the whole volcano is sufficiently cool, a crater lake may occupy the bottom, its water often tainted by fumaroles and thermal springs. The water cannot escape until the cone is dismantled by erosion. Lake Pavir in the Auvergne is an excellent example. In Salvador numerous lakes occupy explosion craters that have been blown in thick deposits of cinerites and pumice; their steep sides, up to 200 m high, cut clearly through these stratified accumulations.

Maare are a particular variety of explosion crater, deriving their name from examples in the Eifel region. At the beginning of the Holocene period and at the end of the Würm glaciation, this region was the scene of much abortive vulcanism, with several basaltic upwellings that only just reached the surface. Explosion craters formed on the site of most of the vents. They

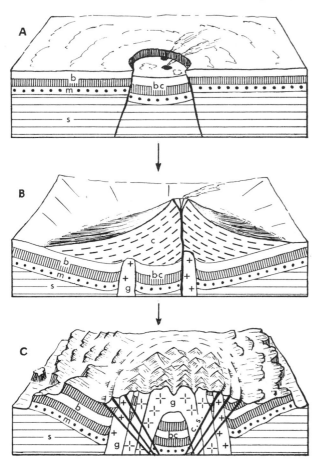

FIG. 4.36. Cauldron subsidence (the 'Scottish' type of
A. Guilcher): the evolution of the caldera on the island of
Mull, in the Inner Hebrides (*from Guilcher, 1950, p. 3*)

In A, the first phase, with basaltic flows (s = basement,
m = Mesozoic, b = basalt). The crater is beginning to
collapse.

In B, the subsidence continues, the early basalt flows
are bent into a syncline, and intrusions of granophyre (g)
occur. A cone builds up over the sunken area.

In C, the present state, after the considerable
dismantling of the cone. The early basalt flows of stage A,
bent into synclinal form, create outward-facing
escarpments; on the site of the former cone there are
plutonic intrusions, and dykes that, radiating outwards,
have a more or less annular outcrop ('ring-dykes')

FIG. 4.37. Cauldron subsidence in the island of Niuafo, in the South Pacific (*after Cotton, 1952, p. 38*)
   A strato-volcano formed of basaltic layers dipping regularly outwards; the centre has collapsed owing to the outflow of the lava, and this has caused a series of step-faults. The subsided area has been invaded by the sea. A small secondary cone has been raised later, after the subsidence

are surrounded by a more or less regular annular ridge, only a few metres high in most cases, formed of scoria and cinders blown out by the explosion. A somewhat similar feature, but much larger, occurs in the Ries district of Swabia, but its origin is in dispute. The Ries was occupied in Oligo-Miocene times by a lake, 20 km in diameter, surrounded by fractures and volcanic vents. It may be partly an explosion crater, that was later enlarged by the formation of a subsidence crater on its margin. But it has also been regarded as the result of the impact of an enormous meteorite.

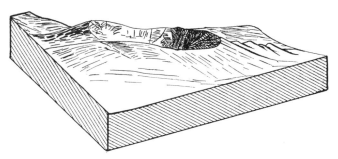

FIG. 4.38. A low cone made by fluid basalt, with a crater much enlarged by an explosion, near San Francisco, Arizona (*after Cotton, 1952, p. 75*)

Indeed, falling meteorites of sufficient dimensions can create holes in the earth's crust rather like gigantic bomb craters. Several are known, in Siberia and the United States in particular. The best criterion for their identification is the presence, on the floor of the crater, beneath the debris that results from the impact, of fragments of the nickel–iron meteorite itself. But it is not always possible to find such evidence, for the meteorite may have burnt up completely as a result of the heat generated by its collision with the earth's surface.

COMPOSITE VOLCANOES
Volcanoes derive their complexity from two factors, on the one hand the interaction between volcanic action and morphogenetic processes, and on

FIG. 4.39. Secondary cone within a caldera, Fogo island (Cape Verde group) (*after Ribeiro, 1954, p. 38*)
  View from the east. The caldera was formed by an explosion, during the first period of activity. During a second episode, a new cone arose, that now slopes directly to the sea

the other hand the varied succession of eruptions of different types. The consequent variety is enormous, and in a sense each example is an individual one. We cannot, therefore, attempt to describe and classify them all, but only to suggest certain guiding lines.

Stratovolcanoes

The term stratovolcano is applied to a cone consisting of more or less regular layers of volcanic ejectamenta. Some authors would insist on the layers being alternations of lavas and pyroclastic materials, but there is no need for such restriction and it seems better to retain a wider connotation; we shall therefore regard a stratovolcano as one formed of piled-up beds, whatever their nature and variety.

  Some volcanoes are produced by eruptions of the Hawaiian type and are thus made up of lava flows that overlap each other. The absence of explosive

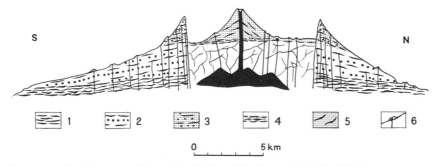

FIG. 4.40. Section across Fogo island (Cape Verde group) (*after Ribeiro, 1954, p. 40*)
  1. Basement lava flows
  2. Alternations of lava and tuff
  3. Tuffs forming the summit of the first cone (now the rim of the caldera)
  4. Lava flows beneath the second cone (within the caldera)
  5. Ash, cinders and lava of the second cone
  6. Adventitious cones of the second paroxysm
The whole structure (see Fig. 4.39) was a stratovolcano, transformed by the first paroxysmal explosion into a caldera; within the caldera a second paroxysm has created another cone and a series of smaller eruption points

action results in the regularity of the bedding. As a result of the fluidity of the lavas, the slopes of the volcano are gentle, of the order of 10°, and usually with a concave profile. The smaller eruptions produce less extensive lava streams that are most numerous around the crater, and this explains the steeper slopes of the upper part of the cone. Some complications may arise from the development of lateral fissures which also exude lava; but these are normally limited to single bulges.

FIG. 4.41. A pyroclastic cone, at the Col de la Coulée de la Soulière, in the Coirons district, France

A small cone formed at the end of the eruptive period that gave the lava flows of the Coirons, and preserving its original shape. Underneath the bushes, the rocks are more coherent, forming the neck. In the foreground, dipping beds of ash, lightly cemented into tuff, shaping the flanks of the volcano

Other stratovolcanoes—those that are so regarded within the stricter meaning of the term—are built up of more or less regular alternations of lava flows, as in the previous case, and more viscous material that now appears in the form of breccias, cinder beds and ashes, and even mud flows produced by lahars. The slopes of the cone are generally steeper and their concavity more marked; on the edge of the crater the breccias and scoria may have a slope of 35°, which is just about the limit for a loose talus. The symmetry of the shapes is less regular than in the case of the Hawaiian type, for the frequent bouts of explosive activity open up fissures around which cones of cinders accumulate, often superimposed upon the more fluid lava streams. Explosion craters may also form, to be partly masked later on by cinder cones, as in the case of the Valle del Bove on the slopes of Etna.

268

There is thus great variety in the relief forms developed on the sides of the cone. The sections exposed in the Valle del Bove show that Etna functioned for a long time under an explosive regime, with the accumulation of breccias, after which there followed a basaltic phase, with the intrusion of basalt dykes into the breccias, and actual lava flows. Later still, lava streams occupied depressions and pyroclastic ejectamenta, covering the

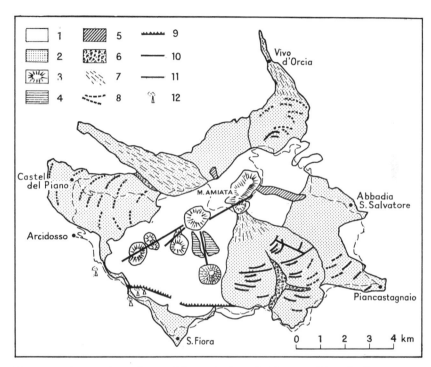

FIG. 4.42. A complex volcano: Monte Amiata, in Tuscany (from Italian geological maps)

   1 and 2. Ignimbrites resulting from fissure-eruptions and explosions (including 'nuées ardentes')
   3. Laccoliths and lava domes, including peaks of peléan type
   4. Flows of glassy lava
   5. Flows of basaltic lava
   6. Explosion breccias
   7. Direction of ignimbrite flows
   8. Transverse ridges on ignimbrite flows
   9. Fractures due to volcanism
10. Pre-existing faults
11. Presumed faults
12. Places where geothermal energy is exploited
An early explosive phase gave rise to an accumulation of ignimbrite flows.
There followed a slight tendency to subsidence, and then the renewal of faulting allowed new eruption centres to develop, with small flows of varying character, and some laccolithic lava domes

almost undissected surfaces of these lavas and accompanied by local cones, indicate a Strombolian regime.

Stratovolcanoes thus offer varied opportunities for the agencies of erosion, and the various layers tend to retain their individuality during the dissection, which is guided by the stratification of the materials. This happens even in the Hawaiian type, for the zone of contact between the superimposed lava flows presents a plane of weakness, just like the stratification planes of thick sedimentary rocks. This zone is often marked by a red layer that represents an ancient soil that has been baked, and by a cindery layer that is easily eroded and so forms a bench on the side of even the steepest gulleys.

The dissection of stratovolcanoes is effected by radial streams that flow down the steepest slopes of the cone. They become progressively incised into the volcanic layers, and as their slope quickly becomes less than the dip of the beds, the various strata are exposed in turn. Their headward erosion may notch the crater rim, taking advantage of zones of weakness, fissures and cinder beds to enlarge the catchment area. The crater thus quickly becomes a watershed, and its walls are carved into roughly triangular promontories that separate true consequent valleys.

The dismantling of the volcano is rendered easier, it is true, when beds of unequal resistance alternate, as in Etna, or better still, in the Cantal, where erosion has proceeded much further since the volcano is of Miocene age. The superimposition of a lava flow on top of unconsolidated breccias or cinder beds gives a hard–soft alternation that produces homoclinal relief forms rather like cuestas though with a much steeper dip. The less dissected lava flows between the major valleys form inclined plateaus—called *planèzes* in the Cantal region—that narrow to a point towards the centre of the volcano. Their pointed ends form sharp flat-iron scarps separating consequent streams. The centre of the volcano is more easily eroded and so becomes a depression at the foot of the *planèzes*. A complete inversion of relief is thus achieved, for differential erosion leaves the plugs, dykes and harder lava flows standing up, whilst the cinder beds and breccias are removed. The ultimate end of such a process is to be seen in the Rhön, in Hesse, where nothing is left but the remains of lava flows that form plateaus, especially where their resistance to erosion is increased by the presence of a resistant substratum. The original architecture of the volcano has thus been completely obliterated.

Complexes of cones and calderas

The term *caldera* is applied to an explosion or subsidence crater of enormous dimensions, usually several kilometres in diameter. It often happens that a caldera cuts across part of a pre-existing cone, and that new cones appear within the caldera. We thus get a complex of cones and calderas, more or less packed together, interlocking or side by side. This type of complex volcanic architecture has been called *Vesuvian*, for Vesuvius provides an excellent example. The Vesuvian type is produced by a series of eruptions emitting scoria, ashes and lava, with occasional and very violent explosive phases. Naturally enough, such an alternation is capable of many variations according to the relative importance of the constructional and explosive periods. The Vesuvian type thus has several subtypes.

Fig. 4.43. The volcano Challapalca, in the Tacna region of Peru
In the foreground, volcanic debris re-deposited by water. The cone, produced by fluid lavas, is very flat, and inversion of relief has commenced; the summit has been dismantled and 'planèzes' are forming

In the case of Vesuvius itself, the main cone was built up as a stratovolcano with alternations of lavas and pyroclastic materials, with trachytes predominating. There followed a period of quiescence during which dissection and soil formation took place. This was interrupted by the eruption of A.D. 79 described by Pliny, that destroyed Pompeii and Herculaneum. This was an extremely violent explosion, with clouds of cinders and ashes, that blew off a portion of the old cone. Such eruptions have been called Plinian in honour of the Latin author. What remains of the rim of the explosion crater now forms Monte Somma, with its steep scarp facing the centre and exposing the layers of the pre-79 stratovolcano. Since that date, activity has continued and has formed, within the explosion crater, the cone that is now known as Vesuvius. An arc-shaped depression, the Atrio del Cavallo, separates Somma from this cone. We thus have a complex consisting of an old stratovolcano, an explosion crater and a new cone.

The formation of calderas is frequent in stratovolcanoes after a period of quiescence. It is often the signal for a new phase of activity that builds up new cones within the caldera, as in Vesuvius. Further—and again Vesuvius provides a type specimen—the caldera is often asymmetrical, having disembowelled a part of the original cone. The island of Santorin, in the Aegean Sea, shows a similar evolution, although in this case the explosion crater has been invaded by the sea. The outer edge of the caldera comprises the even slopes of the original stratovolcano, but in the bay are unstable islands, affected by recent eruptions.

Calderas that are partly due to explosions and partly to subsidence are of frequent occurrence in areas covered by volcanic ash. There is a whole series of them, mostly occupied by lakes, in Salvador, where they occur within an undulating plateau made up of cinerites. Some of the lakes contain small pumice cones. The Flegrean Fields near Naples show closely related phenomena. Here a blanket of cinerites is pitted with subsidence craters that are small but very numerous and sometimes overlapping. Numerous cinder cones are scattered over the area. Calderas are closely associated with pyroclastic deposits, for the explosions which cause them produced masses of ash—as at Krakatao in 1883 and Katmai (Alaska) in 1912.

Subsidence volcanoes (Scottish type)

An unusual evolution has been attributed to the island of Mull in Scotland, and we may call this the *Scottish* type, or 'cauldron subsidence'. It is characterised by stratified beds that dip towards the centre of the structure, rather like a filter-paper in a funnel but of course with a smaller angle of dip. It is more or less completely eroded, and the structure is the opposite of that of an ordinary stratovolcano. It has been produced by depression due to prolonged subsidence. This is not one of the more usual examples of occasional subsidences that tilt this or that area of lava towards the centre of the crater, but a phenomenon sufficiently persistent to give unique characteristics; in other words, it is an extreme case.

On the island of Mull, the emission of mainly basaltic lavas commenced in the Eocene and ended at the beginning of the Oligocene. Subsidence occurred from the very first eruption, with the formation of a lake; and the lavas were poured out under the water. The same thing continued, forming a stratovolcano with a centripetal slope, its centre occupied by a mass of gabbro seamed by divergent basalt dykes. The relief takes on the form of a series of outward-facing cuestas formed by the more resistant lava beds overlying less resistant beds of volcanic conglomerate and breccia. The highest of these scarps is the outermost one, and it attains 300–500 m. In the centre of the areas the relief is more confused, with plugs and domes of lava. The clarity of the relief forms in Mull is largely due to the dissection of what is in fact a very ancient structure. It is probable that the last volcanic accumulations, now removed by erosion, were not so strongly affected by subsidence as the deeper layers that now outcrop round the edges. This is why this type of volcano is an extreme case.

Other subsidence volcanoes exist, in which the deformation is less marked. The most frequent cases, usually associated with stratovolcanoes, show a series of circular cracks on the margins of the crater that have allowed the rocks to be stepfaulted downwards towards the middle. It often happens that subsequent upwellings of magma have filled the cracks, forming ring-dykes, that may ultimately be exposed by weathering. Such annular dykes are found on Mull around the centre of the volcano.

Volcanic structures

The action of weathering agencies is guided by the nature and disposition of the volcanic rocks. Structural factors increase in importance as the volcanic activity declines, and they alone control the subsequent erosion pattern when all is quiet.

Structural volcanic landforms are those that result from the dissection of extinct volcanoes. As with all structural forms, they are controlled by lithology and tectonics.

EVOLUTION OF LAVA FLOWS

In view of the considerable areas occupied by lavas that flowed freely, their geomorphological evolution is a matter of considerable interest. In general, they are characterised by great resistance to erosion, for as we have seen, their thickness, massiveness and the very nature of the basalts of which they are composed, cause them to behave like very hard rocks. Indeed, they resist erosion better than granite under most climates, for they alter more slowly and do not suffer from granular disintegration. Only frost action has much effect, and this attacks the fissured lavas, especially columnar basalts, so that prisms on the edge of the outcrop break and tumble down. A cliff remains, but may crumble under the combined influence of frost and gravity.

Apart from this, the erosional forces have a difficult task. Even the removal of inequalities in the original blocky surface of an *aa* is slow and laboured. Water penetrates too quickly to have much chemical effect; far more effective in provoking mechanical disintegration are the bubbles contained in the lava, for the gases expand and contract in response to temperature changes. Weathered debris accumulates slowly in hollows, and the finer grain of the debris allows more water to accumulate, thus helping chemical action; and gradually the extremely rough microrelief of the *cheire* becomes smoother. In Auvergne, lava flows dating from the middle Quaternary, some 200 000 years ago, which have been through two glacial periods, are now cultivable.

The considerable resistance to erosion offered especially by basalt lava

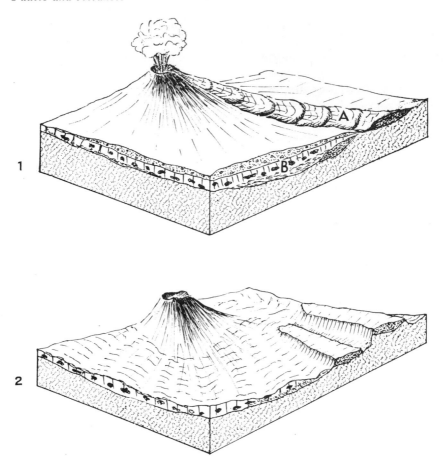

4.44. Inversion of relief in a volcanic area

1. An original volcanic cone with a lava flow (A) and accumulations of pyroclastic materials (B).

2. The lava flow (A) is transformed into a flat ridge, and differential erosion of the lower slopes of the volcano creates a second platform, at a lower level, through the exposure of an old lava flow buried in the pyroclastic material. The cone has been much eroded and only the resistant rocks of the neck remain

flows is responsible for inversions of relief, in which lavas that were laid down in valleys now form hill ridges (Fig. 4.44).

In complex volcanic piles, especially of stratovolcanic type, alternations of lavas and pyroclastic materials may give rise to scarps. Such scarps, of course, do not have the continuity of sedimentary scarps, for the lavas are always highly localised, and both their lithology and their thickness may vary greatly from place to place.

Single lava streams flow down valleys, thus interrupting their geo-morphological evolution, or else spread themselves out over gently undulating or sloping areas. Once the contrast in erosional resistance begins to take effect, inversions of relief begin to appear. Often, as we have seen, water thrown out at the junction of the lava and the country rock starts to form a valley at the side of the lava tongue, and if the country rock is noticeably less resistant than the lava this valley will quickly enlarge, and ultimately the country rock will be worn away completely, leaving the lava flow stranded as a plateau with a fingered end, and scarped sides that slowly recede through the crumbling of the cliffs. Thus a *mesa* is produced, of which a remarkable though complex example is to be found in France in the Coirons. Other smaller and simpler examples are to be seen in the Escandorgne area, near Lodève.

In the Coirons, the Miocene basalt, resulting from fissure eruptions, forms a plateau resting on the crystalline block of the Massif Central. The volcanic rocks comprise many layers of massive basalt, intercalated with beds of scoria and breccia, sometimes with red palaeosols. At Mirabel, basalt cliffs surmount dissected slopes, some 200–300 m high, of Lower Cretaceous clays and limestones. But between the basalt layers are alluvial deposits containing well-rolled granite pebbles; these were left behind by the river whose valley had been occupied by the lava flows. This demonstrates clearly that the lavas were laid down in a depression in the Cretaceous hills, and that the present relief is inverted.

The preservation of such lava flows depends in large measure, however, on the nature of the rocks on which they lie. Like most resistant rocks, the

FIG. 4.45. Basalt mesa at St Jean le Centenier in the Coirons district of France
An abrupt craggy scarp formed by a massive basalt flow overlying the marls and marly limestones of the Lower Cretaceous

basalts play a passive role in the evolution of slopes; their edges recede by the breaking off of large blocks from overhanging cliffs that are undermined by the erosion of the beds underneath. Enormous blocks roll down from time to time on the edges of the Coirons, for the underlying marls are rendered almost fluid by the water filtering down through the lavas. When lavas rest on marls, therefore, the recession of their cliff edges is accelerated. It is not the same when the lava rests on limestone, as is well seen to the south of Privas. In the Rhön area, basalt lava flows have been eroded away when they covered Oligocene marls that were even more susceptible to erosion than the Cretaceous marls of the Coirons, but have remained as mesas where they rest on Muschelkalk limestone which is less easily eroded. The recession of the lava crags thus gradually reduces the extent of the lava flows, giving them first a digitated appearance, well seen on the southern flanks of the Coirons, then reducing them to long fingers, as in the Aubrac, and then to isolated *buttes*.

Somewhat similar to the evolution of lava flows is that of sills or stratified intrusions. In these cases the fluid magma does not reach the surface and so does not create a volcano. Instead it distends the strata and insinuates itself between them, thus forming bedded intrusions. Away from the vent up which the magma wells, there is thus a series of sedimentary beds interlarded with the intrusives. Such a state of affairs can only exist if the sedimentary strata can withstand the intrusion without breaking and continue to offer resistance to the further ascent of the magma. These sills must not

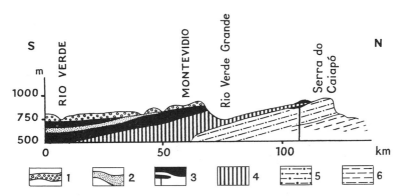

Fig. 4.46. Scarpland relief due to basalt flows in southern Brazil (*after Almeida, 1960, p. 96*)

Basalt flows of Mesozoic age are interbedded in a sandstone formation of varying coarseness and consolidation. Where the dissection is sufficiently mature, the basalts create scarps like the other resistant beds

1. Sands and pebble-beds of the Bauru series
2. Botucatu series, soft sandstone
3. Basalt
4. Soft sandstone with clay partings
5. Hard sandstones
6. Soft pellitic shales

be confused with lava flows that are fossilised underneath subsequent sediments, as in the case of traps. Traps are of frequent occurrence in subsiding areas where the abundant detrital materials succeed in covering up the volcanic lavas. They are well known, for example, in the Trias of New England and Nova Scotia, and in the Permian basin of the Saar-Nahe, in Germany.

From a geomorphological viewpoint there is little difference in the behaviour of sills and that of fossilised lava flows. In both cases denudation is influenced by the stratification of the volcanic rocks and their lithological contrast with the sedimentary rocks in which they occur. In general, since they are formed by fluid lavas, especially basalt, they are harder than most sedimentary rocks and so give rise to scarps and cliffs of limited extent. These conditions are particularly favourable to superimposition, and epigenetic gorges are frequent in areas where sills and trap rocks occur.

DENUDATION OF VENTS, PLUGS AND DYKES

Whereas lava flows and sills have a stratified appearance, the channels by which the magma rose towards the surface are generally vertical. Thus the denudation of these vertical masses of lava does not give rise to plateaus and tabular forms, but to circular or oval features in the case of necks and plugs, or elongated in the case of dykes.

The vertical disposition of these magmatic masses makes them very persistent. Take the case of a volcanic neck: when the volcanic activity

FIG. 4.47. Basalt necks north of Rochemaure, Ardèche, France
Black basalt fills a dyke running parallel to the Rhône, through white Cretaceous marls. Differential erosion has exposed it

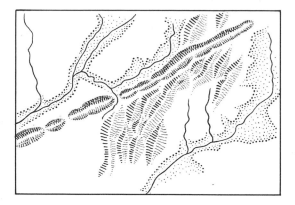

FIG. 4.48. Basic dyke thrown into relief by
differential erosion, south of Marble Bar in Western
Australia (*drawn by Rimbert from an aerial photograph*)

ceases, the magma filling the neck cools and consolidates. This is clearly so
with volcanoes that emit lava, but it is also true of many volcanoes of the
vulcanian type and even more so of the peléan type. It is indeed the
gobbing-up of the throat of the volcano that provokes explosive activity.
Thus it usually happens that the neck consists of a column of resistant
material passing through less consolidated formations of scoria, ashes or
even sedimentary rocks. Such is the case with the *aiguille* that slowly
emerges in a peléan-type eruption. Even though the volcano remains active
in the pelée fashion, the base of the *aiguille* is often laid bare by differential
erosion. The very circumstances of the volcanic process prohibit the

FIG. 4.49. Interstratified tuffs and lavas cut by dykes, on the
inner side of Monte Somma, Vesuvius (*from Cotton, 1952,
p. 228*)
    The softer tuffs are etched out from a chequer-pattern of
dykes and outcrops of harder lavas. Differential weathering
feeds the screes. Situations like this are common in the
vicinity of craters

278

emission of lavas or pyroclastic materials. Thus once the volcanic activity has ceased differential erosion becomes of great importance. The base of an *aiguille* formed of hard and massive rock, with vertical sides, is easily cleared of the unconsolidated material that surrounds it. Part of the base may thus represent an exhumed form but it is usually very difficult to tell exactly how much, for it is almost impossible to say to what height it was engulfed by the unconsolidated debris.

The uncovering of the neck plug may thus go on for a long time, if there is sufficient lithological contrast between it and its enveloping rocks, whether they be volcanic or not. The upper part of the *aiguille* becomes blunted, and its original subaerial form is progressively modified, whilst the base is laid bare. The actual shape of the protrusion scarcely alters during this process, but the relief is sculptured in material that originated at greater and greater depths. Of course, the uncovering of a volcanic neck may also take place, given a sufficient degree of denudation, in the case of volcanoes that do not produce *aiguilles* but only ordinary craters. Such exhumed crater plugs are usually called volcanic necks to distinguish them from genuine *aiguilles*.

While the contents of the vent become a volcanic neck or plug, the dykes may also be exposed. They are generally formed by fluid lavas that make hard rocks. Differential erosion is thus at its best when the dykes traverse unconsolidated pyroclastic materials.

The case of the laccolith is peculiar. Like sills, with which they are generally associated, laccoliths result from magma that rose without reaching the surface. But in this case the intrusion is not a thin layer but a thick domeshaped mass that bulges up the sediments that overlie it. At the surface, the presence of a laccolith shows itself in the form of an anticline, with minor folds perhaps but little fracturing (had the beds been broken the magma would have reached the surface). At depth, in contact with the lava, the structures may be more complex, and some fracturing of the beds may have allowed the magma to rise higher. The laccolith usually forms an irregular series of protuberances that penetrate the overlying rocks, and such bulges are associated with dykes and often incorporate inclusions of sedimentary rock. We may note in conclusion that laccoliths are not only made of lava but may be of plutonic rocks as well, such as granite and rhyolite. Some granite bosses, now largely exposed by denudation, and of circular form with radial and annular dykes, have been regarded as deep-seated laccoliths.

The broad outlines of the geomorphological evolution of laccoliths are relatively simple. In general a radial drainage develops on the dome formed by the sedimentary cover. Naturally enough, this drainage system, as it becomes incised, will be superimposed, and cannot adapt itself to the deeper folds and to the irregular contact between the lava and the overlying rocks. When denudation reaches this contact level, the resulting land forms will be complex in detail. Differential erosion will produce not only epigenetic gorges incised into the lavas that have been stripped of their cover, but also local adaptations to softer layers. When erosion bites more deeply into the structure, things are simpler because the local folds disappear and the rocks are more homogeneous. At a certain level, as in the case of batholiths, there is a magmatic mass of more or less regular structure, with occasional outgrowths and inclusions, surrounded by the rocks into which it was intruded.

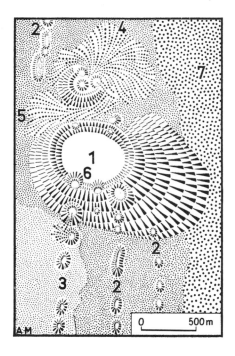

Fig. 4.50. Volcanic alignments at Ludent, in Iceland (*after Thorarinsson, Einarsson and Kjartansson, 1959, p. 159*)
1. Crater of Ludent, more than 9 000 years old
2 and 3. Aligned craters, younger than Ludent
4 and 5. Dacite flows
6. Secondary crater, younger than the alignments
7. Lava flows of about 2 000 years ago
The recurrent character of the volcanic phenomena is apparent

The size of laccoliths is very variable. Some are quite small and are not much larger than the plug of a big volcano. Others, on the contrary, may cover dozens or even hundreds of square kilometres, and may resemble batholiths. The largest laccoliths are often formed of several coalescent masses of magma.

At the conclusion of this review of volcanic landforms, we may draw a parallel with the evolution of structural relief. Tectonic forces modify the configuration of the earth's crust and impose the major structural controls. Whilst they are in action they control morphogenesis through their interference with the external forces. But during periods of quiescence, the forms produced by tectonic forces are more and more modified by erosion, and a pattern of earth sculpture develops, guided by the nature and disposition of the rocks. Renewed dissection, even after a period of planation, also throws into relief the structure and lithology.

Vulcanism is but one manifestation of the internal forces, and it too plays its part in morphogenesis. During active phases, it creates new surface features, either constructional or due to the destructive effect of explosions. The erosion that attacks these features can hardly play more than a very minor role. But once the activity ceases, denudation reduces the volcanic relief to residual forms, and it ends up, as in the case of residual tectonic relief, by sculpturing derived forms the character of which is controlled by

FIG. 4.51. Alignment of basalt cones along a fissure, Craters of the Moon, Idaho, U.S.A. (*after Cotton, 1952, p. 99*)
Steep-sided craters, formed of viscous basalt and cinders, forming chimneys rather than lava-flows

the nature and disposition of the rocks. This influence will be maintained until the volcanic piles have completely disappeared, but it may persist further, in a minor fashion, as a result of the differences in lithology between volcanic plugs and the material surrounding them. These differences only disappear at depth, as for example when the volcanic magma has passed through plutonic rocks. Even then the mechanism of differential erosion may still be maintained, as in the case of basic lavas in the midst of acid plutonic rocks, which behave differently towards chemical weathering.

Bibliography

Chapter 1   The globe

General works

AHRENS, L. H., PRESS, F., RANKAMA, K. and RUNCORN, S. K. (1966) *Physics and Chemistry of the Earth*, Pergamon Press.

'BELOUSSOF, V. V. (1956) 'Grundfragen der allgemeinen Geoteknik', *Geol. Rdsch.* **45**, 353–69.

CAILLEUX, A. (1968) *Anatomy of the Earth*, trans. A. Moody Stuart, London, World University Library.
A good introduction for the general reader. See also Tarling and Tarling below.

CAREY, S. W. 'The asymmetry of the earth', *Aust. J. Sci.* **25**, 369–83 ; 479–88.

FAIRBRIDGE, R. (1961) 'Convergence of evidence on climatic change and Ice Ages', *Ann. N.Y. Acad. Sci.* **95**, 542–79.

FOURMARIER, P. (1962) 'Le problème de l'origine des continents', *Acad. Royale de Belgique, Bull. de la Classe des Sc.* 5ᵉ sér. **48**, 1368–426.

GIDON, P. (1963) *Courants magmatiques et évolution des continents. L'hypothèse d'une érosion sous-crustale*, Paris, Masson (Evol. des Sciences, nº 27).

GOGUEL, J. (1952) *Traité de tectonique*, Paris, Masson.

HILLS, E. S. (1963) *Elements of Structural Geology*, Methuen.

HOLMES, A. (1965) *Principles of Physical Geology*, Nelson.

ROYAL SOCIETY (1965) 'A symposium on continental drift', *Proc. Roy. Soc. London,* **A**, 258, 1–323.

RUNCORN, S. K., ed. (1962) *Continental Drift*, New York, Academic Press.

SITTER, L. U. DE (1959) *Structural Geology*, McGraw-Hill.

SYMPOSIUM (1957) 'Kontinentalaufbau', *Geol. Rdsch.* **46**, no. 1.

TARLING, D. H. and TARLING, M. P. (1971) *Continental Drift*, Bell.

WOOLDRIDGE, S. W. and MORGAN, R. S. (1959). *The Physical Basis of Geography, an Outline of Geomorphology*. Longmans.

The following publications are concerned with more restricted aspects of the subject :

BENIOFF, H. (1954) 'Orogenesis and deep crustal structure. Additional evidence from seismology', *Bull. geol. Soc. Am.* **65**, 385–400.

BERGSTEN, F. (1954) 'The land-uplift in Sweden from the evidence of the old watermarks', *Geogr. Annlr.* **36**, 81–111.

BOURCART, J. (1949) 'La théorie de la flexure continentale', *C. R. Congr. Intern. Géogr., Lisbonne*, **ii**, 167–190.

CHEVALIER, J. M. (1952) 'Analyse harmonique du relief terrestre. Essai d'interprétation mécanique', *Rev. Géom. dyn.* pp. 219–43.

CHEVALIER, J. M. and CAILLEUX, A. (1959) 'Essai de reconstruction géométrique des continents primitifs', *Z. geomorph.* new ser. **3**, 257–68.

DEICHA, G. (1956) 'Aspects géotectoniques possibles de l'équilibre isostatique', *Bull. Soc. géol. Fr.* 6th ser. **6**, 201–8.

FOURMARIER, P. (1954) 'La règle de la compensation approchée des volumes et quelques aspects de la géomorphologie. *Bull. Soc. belge Etud. géogr.* **32**, 23–36.

GILLULY, J. (1949) 'Distribution of mountain building in geologic time', *Bull. geol. Soc. Am.* **60**, 561–90. See discussion in *Geol. Rdsch.* **38** (1950) 89–111.

GLANGEAUD, L. (1962) 'Les transferts d'échelle en géologie et en géophysique. Application à la Méditerranée occidentale et aux chaînes péripacifiques', *Bull. Soc. géol. Fr.* 7th ser. **4**, 912–61.

GOURINARD, Y. (1954) 'Isostasie et déformations quaternaires dans le Nord-Ouest algérien', *C.R. XIXe Congr. Géol. Intern., Alger 1952*, **ix**, 21–38.

GUTENBERG, B. and RICHTER, C. F. (1950) 'Géographie des tremblements de terre et dynamique de la croûte terrestre', *Annls. Inst. Phys. Globe Univ. Strasb.* new ser. **5**, pt 3 (géophysique), 3–11.

JESSEN, O. (1948) *Die Randschwellen der Kontinente*, 2nd edn., Petermanns Mitt. Erg.

KRESTNIKOV, V. N. (1957) 'Séismicité et structure géologique', *Priroda*, no. 8, pp. 25–34.

LAGRULA, J. (1954) 'Sur l'eustatisme et l'isostasie', *C. R. XIXe Congr. Géol. Intern., Alger, 1952*, **ix**, 45–50.

MATSCHINSKI, M. (1951) 'Altitude moyenne des continents et forces géodynamiques', *Rev. Géomorph. dyn.* **2**, 157–65.

MATSCHINSKI, M. (1964) 'Die Erdoberfläche und die Gesetze ihrer Formen', *Z. Geomorph.* new ser. **8**, 163–88.

RUSSO, P. (1953) 'Essai sur les origines de la morphologie terrestre générale', *Rev. Géomorph. dyn.* **4**, 184–200.

SONDER, R. A. (1956) *Mechanik der Erde. Elemente und Studien zur tektonischen Erdgeschichte*, Stuttgart, Schweitzerbart'sche Verlag.
Comprehensive work on the earth's structure and the formation of the continents. Includes references to the major geophysical works by English and German workers.

UMBGROVE, J. H. F. (1951) 'The theory of continental drift', *Adv. Sci.* **8**, 67–70.

Variations in sea level

CAILLEUX, A. (1954) 'Ampleur des régressions glacio-eustatiques', *Bull. Soc. géol. Fr.*, 6th ser. **4**, 243–54.

CASTANY, G. and OTTMANN, F. (1957) 'Le Quaternaire marin de la Méditerranée occidentale', *Rev. Géogr. phys. Géol. dyn.*, 2nd ser. **i**, 46–55.

CHOUBERT, G (1962) 'Réflexion sur les parallélismes probables des formations quaternaires atlantiques du Maroc avec celles de la Méditerranée', *Quaternaria*, **6**, 137–75.

DONN, W., FARRAND, W. and EWING, M. (1962) 'Pleistocene ice volumes and sealevel lowering', *J. Geol.* **70**, 206–14.

FAIRBRIDGE, R. (1958) 'Dating the latest movements of the quaternary sealevel', *Trans. N.Y. Acad. Sci.* 2nd ser. **20**, 471–82.

GIGOUT, M. (1956) 'Réponse au questionnaire de la Commission des lignes de rivage du IVᵉ Congrès International de l'INQUA (1953)', *Quaternaria*, **3**, 71–9.

GRAUL, H. (1959) 'Der Verlauf des glazialeustatischen Meeresspielgelanstieges, berechnet an Hand von C_{14} Datierungen', *Deutscher Geographentag*, Berlin, pp. 232–42.

GUTENBERG, B. (1941) 'Changes in sea-level, postglacial uplift, and mobility of the Earth's interior', *Bull. geol. Soc. Am.* **52**, 721–72.

KUENEN, PH. (1954) 'Eustatic changes of sea-level', *Geologie Mijn.* new ser. **16**, 148–55.

KUENEN, PH. (1955) 'Sea-level and crustal warping', *Geol. Soc. Am.* special paper no. 62, pp. 193–204.

SYMPOSIUM (1954) 'Quaternary changes in sea-level, especially in the Netherlands', *Geol. mijnbouw*, new ser. **16**, 147–70.

TRICART, J. (1958) 'Mise au point: les variations quaternaires du niveau marin', *Inf. géogr.* pp. 100–5.

UNION GÉOGRAPHIQUE INTERNATIONAL (1956) *Premier rapport de la commission pour l'étude et la corrélation des niveaux d'érosion et des surfaces d'aplanissement autour de l'Atlantique.* Union Géogr. Intern. 5 vols.

WEGMANN, E. (1955) 'Tectonique vivante: vue d'ensemble sur les travaux de la réunion de printemps 1954 à Mayence', *Geol. Rdsch.* **43**, no. 1, 273–306.

WOLDSTEDT, P. (1954) *Das Eiszeitalter. Grundlinien einer Geologie des Quartärs.* vol. 1. *Die allgemeinen Erscheinungen des Eiszeitalters*, 2nd edn., Stuttgart, Enke.

Chapter 2 Geosynclines and fold belts

While the publications on the subject of structural geology are almost innumerable, most pay scant attention to morphological expression. An introduction to some aspects covered in this chapter may be found in:

TWIDALE, C. R. (1971) *Structural Landforms*, New York, M.I.T. Press.

For a fuller bibliography see:

TRICART, J. and CAILLEUX, A. M. (1962) *Traité de Gémorphologie.*

Geosynclines

AUBOUIN, J. (1958) 'A propos d'un centenaire: les aventures de la notion de géosynclinal', *Rev. Géogr. phys. Géol. dyn.* new ser. **2**, 135–88.

AUBOUIN, J. (1961a) 'Propos sur les géosynclinaux', *Bull. Soc. géol. Fr.* 7th ser. **3**, 629–702.

DZULINSKI, S., KSIAZKIEWICZ and KUENEN, P. (1959) 'Turbidites in flysch of the Polish Carpathian Mountains', *Bull. Geol. Soc. Am.* **70**, 1089–118.

FIELD, R. M. (1952) Geophysical-geochemical-geological significance of geosynclines. *C.R. XIXe Congr. Intern. Géol., Alger,* **xiv**, 181–99.

FLANDRIN, J. (1955) 'Remarques sur quelques hypothèses récentes de tectonique algérienne', *Bull. Soc. géol. Fr.,* 6th ser. **5**, 391–98.

FOURMARIER, P. (1948) 'Asymétrie structurale des tectogènes', *Ann. Soc. Géol. belg.* **71** B, 179–93.

GILLULY, STILLE, KREJCI-GRAF, WEGMANN (1950) 'Struktur und Zeit', *Geol. Rdsch.* **38**, no. 2, 89–132.

GLAESSNER and TEICHERT (1947) 'Geosynclines: a fundamental concept in geology', *Am. J. Sci.* **245**, 465–82, and 571–91.

GLANGEAUD, L. (1955) 'Les déformations plio-quaternaires de l'Afrique du Nord', *Geol. Rdsch.* **43**, 181–96.

KUENEN, P. (1959) 'Turbidity currents, a major factor in flysch deposition', *Ecloq. geol. Helv.* **51**, 1009–21.

LUCIUS, M. (1952) 'L'évolution des conceptions sur la genèse et le mécanisme des plissements de l'écorce terrestre., *Bull. Soc. Nat. luxemb.* new ser. **46**, 64–121.

MOURATOV, M. (1962) 'Histoire de l'évolution tectonique de la zone plissée alpine de l'Europe orientale et de l'Asie Mineure', *Bull. Soc. géol. Fr.* 7th ser. **4**, 182–200.

VEYRET, P. (1957a) 'Nouveautés sur la structure des Alpes', *Rev. Géogr. alp.* **45**, 215–50.

On the general nature of mountain chains and the relationships with structure and drainage patterns

BEMMELEN, R. VAN (1955) 'L'évolution orogénétique de la Sonde (Indonésie)', *Bull. Soc. Belge Géol. Hydrol.* **64**, 124–52.

DRESCH, J. (1950) 'Questions morphologiques des hautes plaines Constantinoises', *Bull. Ass. Géogr. fr.* pp. 90–5.

GABERT, P. (1960) 'Une tentative d'évaluation du travail de l'érosion sur les massifs montagneux qui dominent la Plaine du Pô', *Rev. Géogr. alp.* **48**, 593–605.

GLANGEAUD, L. (1953) 'Tectonique comparée des nappes de glissement dans le Jura bressan et diverses régions méditerranéennes', *Bull. Soc. géol. Fr.* 6th ser. **7**, 697–702.

GLANGEAUD, L. (1957) 'Essai de classification géodynamique des chaînes et des phénomènes orogéniques', *Rev. Géogr. phys. Géol. dyn.* new ser. **1**, 200–20.

MASSEPORT, J. (1955) 'Le Sillon Alpin, dépression d'érosion ou déchirure structurale?' *Rev. Géogr. alp.* **43**, 793–820.

RODGERS, J. (1971) *The Tectonics of the Appalachians,* Wiley/Interscience.

VEYRET, P. (1955) 'Le réseau hydrographique de la Chartreuse et du Vercors (Préalpes du Nord) à la lumière des idées tectoniques nouvelles', *Rev. Géogr. alp.* **43**, 697–702.

VEYRET, P. (1956) 'La cluse de Grenoble: contribution à l'étude du relief plissé', *Rev. Géogr. alp.* **44**, 297–310.

VEYRET, P. (1957b) 'Le problème des cluses préalpines: la Cluse de Chambéry', *Rev. Géogr. alp.* **45**, 9–28.

VIERS, G. (1956a) 'L'évolution du réseau hydrographique dans le Pays basque français', *Pirineos*, **12**, 159–89.

Overthrust nappe regions

AUBOUIN, J. (1961b) 'Contribution à l'étude géologique de la Grèce septentrionale: les confins de l'Epire et de la Thessalie', *Ann. géol. Pays hell.* **10**

BORDET, P. (1961) '*Recherches géologiques dans l'Himalaya du Népal, région de Makalu*', Paris, Centre Nationale des Recherches Scientifiques.

CAIRE, A. (1962) 'Phénomènes tectoniques de biseautage et de rabotage dans le Tell algérien', *Rev. Géogr. phys. et Géol. dyn.*, 2nd ser. **5**, 299–325.

CAIRE, A. (1957) *Etude géologique de la chaîne des Bibans (Algérie),* Bull. *Serv. Carte Géol. Algérie,* new ser. no. 16, 2 vol.

CAIRE, A., CHAUVE, P., GLANGEAUD, L. and MATTAUER, M. (1955) 'Sur la structure d'une partie de l'Atlas tellien septentrional', *Bull. Soc. géol. Fr.* 6th ser. **5**, 267–78.

CAIRE, A., GLANGEAUD, L. and MATTAUER, M. (1951) 'L'évolution structurale de la région de Miliana (Algérie) et le style amygdalaire des chaînes liminaires', *Bull. Soc. géol. Fr.* 5th ser. **20**, 479–502.

GAGNEBIN, E. (1945) 'Quelques problèmes de la tectonique d'écoulement en Suisse orientale', *Bull. Lab. Géol. Univ. Lausanne,* no. 80.

GEZE, B. (1963a) 'Caractères structuraux de l'Arc de Nice, Alpes-Maritimes', *Livre Mém. P. Fallot,* Soc. Géol. Fr. **ii**, 289–300.

KLIMASZEWSKI, M. (1956) 'Le développement des Carpates occidentales', *Priroda*, no. 7, pp. 68–73 (in Russian).

LEMOINE, M. (1959) 'Un exemple de tectonique chaotique: Timor. Essai de coordination et d'interprétation', *Rev. Géogr. phys. et Géol. dyn.* 2nd ser. **2**, 205–30.

SITTER, L. U. DE (1954) 'Gravitational gliding tectonics: an essay in comparative structural geology', *Am. J. Sci.* **252**, 321–44.

Cordilleras and *plis de fond*

BEHRMANN, R. B. (1958) 'Die geotektonische Entwicklung des Apennin-Systems', *Geotektonische Forsch.* no. 12, 99 pp.

BIROT, P. (1960a) 'Esquisse morphologique des Monts Celtibériques orientaux', *Bull. Sect., Géogr. Comité Trav. Hist. Sci.* Min. Ed. Nat., **72**, (1959), pp. 101–30.

CASTERAS, M. (1952) 'Esquisse structurale du versant Nord des Pyrénées', *C.R. XIXe Congr. Géol. Intern., Alger.* **xiv**, 165–80.

DOLLFUS, O. (1959/60) 'Présentation de la structure des Andes Centrales péruviennes', *Trav. Inst. Fr. Etudes Andines,* **7**, 53–64.

DRESCH, J. (1941) 'Recherches sur l'évolution du relief dans le Massif Central du Grand Atlas, le Haouz et le Sous', Arrault, Tours.

FEL, A. (1962) 'Les montagnes de Ligurie (essai sur la formation d'un relief)', *Rev. Géogr. alp.* **50**, 311–78.

GERTH, H. (1955) *Bau der südamerikanischen Kordillere.* Berlin, Bornträger.

JOLY, F. (1962) *Etudes sur le relief du Sud-Est marocain.* Trav. Inst. Scient. Chérifien. Sér. Géol., no. 10.

MCBIRNEY, A. R. (1963) 'Geology of a part of the Central Guatemalan Cordillera', *Calif. Univ. Public. Geol. Sc.* **38**, 177–242.

MENNESSIER, G. (1961) 'Les caractères structuraux des montagnes de la région de Kaboul (Afghanistan)', *Bull. Soc. géol. Fr.,* 7th ser. **3**, 40–9.

PROUST, F. (1962) 'Tectonique de socle par failles inverses en liaison avec d'anciennes failles normales, dans le Haut-Atlas (Maroc)', *C.r. Somm. Soc. géol. Fr.* pp. 9–11.

RAYNAL, R. (1961) 'Plaines et piedmonts du bassin de la Moulouya (Maroc oriental). Etude géomorphologique', Rabat.

VIERS, G. (1960a) 'Nouvelles conceptions touchant l'évolution géologique des Pyrénées', *Rev. Géogr. Pyrénées S-Ouest,* **31**, 85–92.

WEBER, C. (1961) 'Structure géologique de la Californie: revue des problèmes actuels', *Rev. Géogr. Phys. et Géol. Dyn.* new ser. **4**, 67–88.

WINKLER VON HERMADEN, A. (1960) 'Zur Frage der Beziehungen zwischen Tektonik und Landformung', *Geol. Rdsch,* **50**, 273–90.

Intramontane basins

Some works already cited may be consulted on this subject, notably those by: R. Raynal (1961), J. Dresch (1941) and Mouratov (1962).

BIROT, P. and SOLE SABARIS, L. (1959) Recherches sur la morphologie du Sud-Est de l'Espagne', *Rev. Géogr. Pyr. S-Ouest,* **30**, 209–84.

BOMER, B. (1956a) 'Aspects morphologiques du bassin de Catalayud-Daroca et de ses bordures', *Bull. Ass. Géogr. fr.* **261/2**, 186–94.

CASTANY, G. (1948) 'Les fosses d'effondrement de Tunisie: géologie et hydrologie', *Ann. Mines et Géol., Tunis*, no. 3, 126 pp.

PANNEKOEK, A. (1960) 'Post-orogenic history of mountain-ranges', *Geol. Rdsch.* **50**, 259–73.

RAYNAL, R. (1952) 'La région de la Haute-Moulouya', *Serv. Géol. Maroc, Notes et Mém.* **96**, 53–69.

RONAI, A. (1963) 'Etude sur les couches fluviatiles dans le grand bassin hongrois', *INQUA, 6ᵉ Congr., Warsaw 1961, Reports*, **iii**, 301–7.

TRICART, J. and MICHEL, M. (1964) 'La geomorfologia de la cuenca de Santiago y sus relaciones con las aguas subterràneas', *Bull. Fac. Lettres Strasb.* pp. 446–60.

Folding in plastic cover rocks

Works cited are essentially those with a geomorphological rather than a geological approach. See also R. Coque (1962) and R. Raynal (1961).

BARRERE, P. (1962) 'Reliefs mûrs perchés de la Navarre orientale', *Rev. Géogr. Pyrénées et S.O.* **33**, 309–23.

BRUNET, P. (1958) 'Recherches morphologiques sur les Corbières', *Mém. et Doc. Centre Document. Cartogr. CNRS*, **6**, 59–124.

CLAUZON, G. (1961) 'La chaîne de l'Etoile', *Bull. Ass. Géogr. fr.* **298**, 74–82.

GLANGEAUD, L. (1944) 'Le rôle des failles dans la structure du Jura externe', *Bull. Soc. Hist. Nat. Doubs.* **51**, 17–38.

GLANGEAUD, L. (1949) 'Les caractères structuraux du Jura', *Bull. Soc. géol. Fr.* 5th ser. **19**, 669–88.

GOGUEL, J. (1948) 'Le rôle des failles de décrochement dans le Massif de la Grande Chartreuse', *Bull. Sc. géol. Fr.* 5th ser. **18**, 227–35.

GOHAU, G. and VESLIN, J. (1958) 'Un exemple de morphotectonique en Haute-Provence (Pays niçois)', *Rev. Géogr. phys. et Géol. dyn.* new ser. **2**, 189–92.

LEBEAU, R. (1951) 'Sur la structure du Jura: les enseignements de l'excursion géologique interuniversitaire de Franche-Comté (31 Août–6 Sept. 1949)', *Rev. Géogr. Lyon*, **26**, no. 1.

POUQUET, J. (1951) *Les Monts du Tessala (chaînes sud-tellinenns d'Oranie), essai morphologique*, Paris, SEDES.

VEYRET, P. (1960a) 'Le problème de l'inversion du relief en Chartreuse (et dans les Préalpes françaises du Nord)', *Rev. Géogr. alp.* **48**, 227–66.

VEYRET, P. (1960b) 'L'inversion de relief en Chartreuse, précisions et compléments', *Rev. Géogr. alp.* **48**, 585–92.

VIERS, G. (1956b) 'Les synclinaux perchés de la Haute-Sioule et leur modelé', *Rev. Géogr. Pyr. S.O.* **27**, 329–57.

VIERS, G. (1960) *Le relief des Pyrénées occidentales et de leur piémont. (Pays Basque français et Barétous)*, Toulouse, 604 pp., 85 figs.

Folding in rigid cover rocks

Reference should also be made to the following works already cited: J. Dresch (1941), F. Joly (1962), R. Raynal (1961), O. Dollfus (1959/60), R. Coque (1962).

BOMER, B. (1954a) 'Trois aspects du contact entre Monts Deltibériques Occidentaux et Bassin de l'Ebre', *Bull. Ass. Géogr. fr.* pp. 35–41.

DRESCH, J. (1952) 'Le Haut-Atlas occidental', *Serv. Géol. du Maroc, Notes et Mém.* **96**, 107–21.

FLANDRIN, J. (1952) 'La Chaîne du Djurdjura', *XIXᵉ Congr. Géol. Intern., Alger, Monogr. Région,* 1st ser. no. 19, 50 pp.

FONTBOTE, J. M. (1954) 'Las relaciones tectônicas de la depresiôn del Vallés-Panedés con la Cordillera prelitoral catalana y con la depresiôn del Ebro', *Bol. Soc. Esp. Hist. Nat.* pp. 281–310.

PLANHOL, X. DE (1956) 'Contribution à l'étude géomorphologique du Taurus occidental et de ses plaines bordières', *Rev. Géogr. alp.* **44**, 86 pp.

VAUMAS, E. DE (1954) *Le Liban. Etude de géographie physique,* Paris, Firmin-Didot, 367 pp.

VAUMAS, E. DE (1957) 'Plateaux, plaines et dépressions de la Syrie intérieure septentrionale. Etude morphologique', *Bull. Soc. Géogr. Egypt,* **30**, 97–236.

Diapirs

JUNG, J. and SCHLUMBERGER, C. (1936) 'Soulèvement, des alluvions du Rhin par les intrusions salines diapires de la Haute-Alsace', *Bull. Serv. Carte Géol. Als.-Lorr.* **3**, 77–86.

MÜCKE, E. (1959) 'Entwicklungsgang und Formenbildung der Salzauslaugung in N.O. Thüringen', *Wiss. Z. Martin-Luther-Univ. Halle-Wittenb., Math-Nat.* **8**, 641–50.

PEVNEV, A. K. (1962) 'Vertical crustal movements of the Baskunchak Salt Dome', *Abh. Dtschen. Ak. Wiss. Berlin,* **i**, *Intern. Symp. Erdkrustenbeweg,* pp. 81–7.

PICARD, K. (1958) 'Die Struktur Peissen und das glazigene Geschehen im Raum des Lockstedter Sandes', *Geol. Jahrbuch,* **75**, 347–58.

ROLL, A. (1956) 'Zur Strukturgeschichte der Salzstöcke von Wesendorf und Hohenhorn', *Geotektonisches Symp. H. Stille,* pp. 228–45.

Chapter 3　The platform regions

There is an enormous literature on the subjects dealt with in this chapter. Few of these are comprehensive, geosynclincs having dominated a great deal of work. As for the fold chains, the works cited are predominantly by French and Russian scientists who have advanced our knowledge of the structural geomorphology of the platforms. This explains the choice of examples and references.

General works

Books on tectonics, strucure and geology by Goguel, Hills and Sonder should also be consulted (see p. 282–3).

Tectodynamics

BELOUSSOV, V. V. (1960) 'Tectonic map of the earth', *Geol. Rdsch.* **50**, 316–24.

DREYFUS, M. (1960) 'Déformations contemporaines de la sédimentation dans le Jura', *Bull. Trimest. Serv. Inf. géol. Bur. Rech. géol. geophys. et minières*, **12**, 1–7 (no. 47).

MESHCHERIAKOV, IOU. A. (1959) 'Secular crustal movements of the East European plain and associated problems', *Bull. Géodésique*, **52**, 69–75.

Granitisation

BIROT, P. (1960b) 'Les reliefs résiduels des socles cristallins', *XVIIIᵉ Congr. Intern. Géogr., Rio de Janeiro, 1956*, **ii**, 151–3.

BOISSONNAS, J. (1963) 'Les structures annulaires des granites des « taourirts » en Ahaggar occidental (Sahara)', *Bull. Soc. géol. Fr.* 7th ser. **5**, 695–700.

COGNE, J. (1960) 'Métamorphisme et granitisation en liaison avec l'évolution orogénique en Bretagne méridionale', *Bull. Soc. géol. Fr.* 7th ser. **2**, 213–26.

HOUGHTON-BRUNN, J. (1961) 'Origine et localisation de l'énergie de la granitisation', *C.R. Acad. Sci.* **252**, 3470–2.

ROCH, E. (1958) 'Granites laccolithiques, granites ambatomiranty et granites stratoïdes de Madagascar', *C.r. Somm. Soc. géol. Fr.* pp. 130–3.

Etudes régionales

AB'SABER, A. NACIB (1956) 'A terra paulista', *Bol. paul. Geogr.* **23**, 5–38.

BOGDANOFF, A. (1957) 'Traits fondamentaux de la tectonique de l'URSS', *Rev. Géogr. phys. et Géol. dyn.* 2nd ser. **1**, 135ff.

BORISOV, A. A. (1963) 'Some deep-seated features of platform provinces of the Soviet Union', *Internat. Geol. Rev.* **5**, 815–29.

ELHAÏ, H. (1963) 'La Normandie occidentale entre la Seine et le Golfe Normanno-Breton, étude morphologique', Thesis, Paris.

HILLS, E. S. (1956) 'The tectonic style of Australia', in *Geotektonisches Symposium zu Ehren von Hans Stille*, Stuttgart.

LEPERSONNE, J. (1956a) 'Les surfaces d'érosion des hauts plateaux de l'intérieur de l'Afrique Centrale', *Bull. Séances Ac. Sc. Col., Bruxelles*, **2**, 596–621.

LEPERSONNE, J. (1960) 'Quelques problèmes de l'histoire géologique de l'Afrique au Sud du Sahara depuis le Carbonifère', *Bull. Soc. géol. belg.* **84**, 21–85.

SUSLOV, S. P. (1961) *Physical Geography of Asiatic Russia*, trans. N. Gershevski and J.-E. Williams. New York and London, Freeman.

VAUMAS, E. DE (1961) 'Structure et morphologie du Proche-Orient. Nouvel essai de synthèse et orientations de recherche', *Rev. Géogr. alp.* **49**, 433–509.

WELLINGTON, J. H. (1946) 'A physiographic regional classification of South Africa', *South Afr. Geogr. Journ.*, **28**, 64–86.

WELLINGTON, J. H. (1955) *Southern Africa, a Geographical Study*, Cambridge Univ. Press, vol. i, pp. 11–125.

Cratonic arches

ALMEIDA, F. DE (1953) 'Considerações sôbre a geomorfogênese da Serra do Cubatào', *Bol. paul. Geogr.* **15**, 3–17.

BIGARELLA, J. J. and AB'SABER, A. N. (1964) 'Paläogeographische und Paläoklimatische Aspekte des Känozoikums in Südbrasilien', *Z. Geomorph*, new ser. **8**, 286–312.

BIROT, P. (1957) 'Esquisse morphologique de la région littorale de l'état de Rio de Janeiro', *Ann. de Géogr.* **66**, 80–91.

DRESCH, J. (1957) 'Les problèmes morphologiques du Nord-Est brésilien', *Bull. Ass. Géogr. fr.* pp. 48–59.

FAIR, T. and KING, L. C. (1954) 'Erosional land-surfaces in the eastern marginal areas of South Africa', *Trans. geol. Soc. S. Afr.* **57**, 19–26.

KING, L. C. (1956) 'Drakensberg scarp of South Africa: a clarification', *Bull. geol. Soc. Am.* **67**, 121–2.

KING, L. C. (1959) 'Denudational and tectonic relief in South-Eastern Australia', *Trans. geol. Soc. S. Afr.* **62**, 113–38.

MAUGIS, P. (1955) 'Reconnaissance pétrolière (géologique et géophysique) du bassin sédimentaire de la Côte d'Ivoire', *Bull. D.F.M.G., Dakar*, no. 19, pp. 11–95.

VANN, J. H. (1959) 'The geomorphology of the Guiana coast', *Second Coastal Geogr. Conf., Bâton-Rouge*, pp. 155–87.

Shields

BIAYS, P. (1960) 'Quelques travaux et documents concernant le Bouclier canadien', *Norois*, **7**, 13–31.

CHOUBERT, B. (1960) 'Le problème des structures tectoniques surimposées en Guyane française', *Bull. Soc. géol. Fr.* 7th ser. **2**, 855–61.

EVERS, W. (1960) 'Die Geomorphologie der Skanden', *Petermanns Mitt.* pp. 89–102.

FENELON, P. (1957) 'La plaine à inselbergs de Patos (Etat de Paraiba, Brésil)', *Bull. Ass. Géogr. fr.* pp. 60–5.

MULCAHY, M. J. and HINGSTON, F. J. (1961) *The Development and Distribution of the Soils of the York–Quairading area, Western Australia, in relation to Landscape Evolution.* CSIRO, Soil Publ. no. 17.

RUELLAN, F. (1952) 'Le rôle des plis de fond dans la structure et le relief du bouclier sud-américain', *Congr. Intern. Géol., Alger*, **iii**, 241–61.

TANNER, V. (1938) *Die Oberflächengestaltung Finnlands. Eine übersichtliche Darstellung der Morphographie und Morphologie sowie der Morphogenie in chronologischer Beziehung.*

WOLFF, C. W. and SWARZENSKI, W. (1960) 'The tectonic significance of the erosion surfaces in North-Western Maine', *Z. Geomorph.* new ser. **4**, 53–68.

Appalachian relief

BETHUNE, P. DE (1948) 'Geomorphic studies in the Appalachians of Pennsylvania', *Am. J. Sci.* **246**, 1–22.

BETHUNE, P. DE (1957) 'Le relief du Condroz', *Tijds. Kon. Nederl. Aardrijks. Gen.* **74**, 220–33.

DIBNER, V. D. (1957) 'Au sujet de la genèse du relief de l'Oural', *Izv. Vsiess. Geogr. Obchtch.* pp. 131–7 (in Russian).

DRESCH, J. (1953) 'Morphologie de la Chaîne d'Ougarta', *Trav. Inst. Rech. Sahariennes*, **9**, 25–38.

LLOPIS LLADO, N. and FONTBOTE, J. (1959) *Estudio geologico de la Cabrera Alta (León)*, Depart. Geogr. Aplicada Zaragoza.

Plateaux

BIROT, P., CAPOT-REY, R. and DRESCH, J. (1955) 'Recherches morphologiques dans le Sahara Central', *Trav. Inst. Recherches Sahariennes*, **13**, 13–74.

DAVEAU, S. (1959) 'Recherches morphologiques sur la région de Bandiagara', *Mém. I.F.A.N.* no. 56.

DAVEAU, S. (1960) 'Les plateaux du Sud-Ouest de la Haute-Volta, étude géomorphologique', *Trav. Dép. Géogr. Univ. Dakar*, **7**.

DRESCH, J. (1961) 'Plaines et « eglab » de Mauritanie', *Bull. Ass. Géogr. fr.* **299/300**, 130–42.

PERRY, R., MABBUTT, J. et al. (1962) *Lands of Alice Springs Area, Northern Territory, 1956–57.* CSIRO, Land Res. Ser. no. 6.

Ancient massifs

BEAUJEU-GARNIER, J. (1950) *Le Morvan et sa bordure, étude morphologique,* Paris, P.U.F.

BEAUJEU-GARNIER, J. (1952) 'Quelques données nouvelles à propos des massifs anciens', *Rev. Géomorph. dyn.* **3**, 57–77.

BEAUJEU-GARNIER, J. (1954) 'Essai de morphologie limousine', *Rev. Géogr. alp.* pp. 269–302.

BERTRAND, G. (1960) 'Traits morphologiques originaux du plateau de Montredon-Labessonnière (SW du Massif Central)', *Rev. Géogr. Pyrénées et S.O.* **31**, 277–93.

BLENK, M. (1960) *Morphologie des nordwestlichen Harzes und seines Vorland.* Göttinger Geogr. Abh. no. 24.

BOMER, B. (1954b) 'Le relief du Limousin septentrional', *Mém. Centre Doc. Cartogr. C.N.R.S.* **4**, 63–95.

COULET, E. (1952) 'Contribution à l'étude morphologique de la Margeride', *Bull. Soc. Languedocienne de Géogr.* **24**, 201–334.

DAUTRY, J. (1958) 'Les reliefs résiduels de l'Ouest Montluçonnais', *Norois*, **5**, no. 18, 139–50.

FEIO, M. (1952) *A evoluçao de relêvo do Baixo Alentejo, estudio de geomorfologia*, Centro de Est. Geogr. Inst. para a alta cultura, Lisbonne.

FRANZ, H. J. (1956) 'Die Oberflächengestaltung des Harzes. *Z. Erdk-Unterr.* **8**, 1–13.

GAMBLIN, A. (1954) 'Quelques aspects de la morphologie de la Fagne de Chimay', *Bull. Ass. Géogr. fr.* 68–78.

GELLERT, J. (1962) 'Morphogenetische Untersuchungen in den sächsich-thüringischen Mittelgebirgen', *Geogr. Ber.* no. **23**, 209–14.

HERMANS, W. F. (1955) *Description et gnèse des dépôts meubles de surface et du relief de l'Oesling.* Serv. Géol. Luxembourg.

LUCIUS, M. (1957) 'L'évolution du relief de l'Oesling (Ardennes luxembourgeoises)', *Tijds. K. ned. aardrijsk. Genod.* **74**, 313–23.

MACAR, P. (1954) 'L'évolution géomorphologique de l'Ardenne', *Bull. Soc. roy. belge Géogr.* **78**, 23 pp.

MACAR, P. and ALEXANDRE, J. (1960) 'Pénéplaine unique plio-pléistocène et couverture tertiaire ayant noyé les dépressions préexistantes en Haute-Belgique', *Bull. Soc. Belge Géol. Paléont. Hydrol.* **69**, 295–315.

PELLETIER, J. (1953) 'La bordure orientale du Massif Central de Vienne à Tournon, étude morphologique', *Rev. Géogr. Lyon*, **28**, 359–69.

PISSART, A. (1961/2) 'Les aplanissements tertiaires et les surfaces d'érosion anciennes de l'Ardenne du Sud-Ouest', *Annls Soc. géol. Belg.* **85**, Mém. no. 2, pp. 71–150.

SANDNER, G. (1956) 'Der Kellerwald und seine Umrahmung. Eine geomorphologische Untersuchung', *Marburger geogr. Schriften*, no. 4.

SOONS, J. (1958) 'Landscape evolution in the Ochil Hills', *Scottish Geogr. Mag.* **74**, 86–97.

TERS, M. (1961) 'La Vendée Littorale, étude de géomorphologie', Thesis, Paris.

VOGT, J. (1962a) 'A propos de la morphologie des confins du Limousin et du Perigord', *Rev. Géogr. alp.* pp. 121–6.

WEBER, H. (1956) 'Formenkundliche Probleme im Thüringischen Gebirge', *Hallesches Jahrb. Mitteldeutsche Erdgeschichte*, **2**, no. 3, 142–65.

293

Sedimentary basins

ALBAGNAC, J. (1954) 'La région entre l'Eure et l'Iton, étude morphologique', *Bull. Ass. Géogr. Fr.* pp. 15–25.

BROWN, E. (1960) 'The building of southern Britain', *Z. Geomorph.* new ser. **4**, 264–74.

BÜDEL, J. (1957) 'Grundzüge der klimamorphologischen Entwicklung Frankens', *Würzburger Geogr. Arbeiten*, **4/5**, 5–46.

CHOLLEY, A. (1953) 'Quelques aperçus nouveaux sur l'évolution morphologique du Bassin de Paris', *Ann. de Géogr.* **62**, 4–17, 92–107.

CHOLLEY, A. (1960) 'Remarques sur la structure et l'évolution morphologique du Bassin Parisien', *Bull. Ass. Géogr. Fr.* **288/9**, 2–25.

ENJALBERT, H. (1960) *Les Pays aquitains: le modelé et les sols.* Bordeaux, Bière.

GRAS, J. (1963) 'Le Bassin de Paris méridional, étude morphologique', Thesis, Paris.

KEGEL, W. (1957) 'Das Parnaiba-Becken', *Geol. Rdsch.* pp. 522–39.

MATHIEU, G. and FACON, R. (1953) 'Relations entre la morphologie et la structure du seuil du Poitou', *C.R. Acad. Sci.* **236**, 1982–4.

PINCHEMEL, P. (1953) 'Introduction morphologique à la géographie du Boulonnais et du Weald', *Rev. du Nord*, **35**, 29–46.

TRICART, J. (1949) *La Partie orientale du Bassin de Paris, étude morphologique.* 2 vols, Paris, SEDES.

WOOLDRIDGE, S. W. and LINTON, D. L. (1955) *Structure, Surface and Drainage in South-East England.* London, Philip.

Cuesta landforms

AHNERT, F. (1960) 'The influence of Pleistocene climates upon the morphology of cuesta scarps on the Colorado Plateau', *Ann. Ass. Am. Geogr.* **50**, 139–56.

ALMEIDA, F. (1960) 'O Planalto basáltico da bacia do Paraná, América do Sul', *XVIII Congr. Intern. Géogr., Rio de Janeiro, 1956* **ii**, 91–112.

AWAD, H. (1957) 'A new type of desert cuesta in Central Sinaï', *Proc. XVIIth Geogr. Congr., Washington, 1952* **i**, 126–31.

BLUME, H. (1960) 'Probleme der Stufenlandschaft erläutert am Beispiel des Luxemburger Gutlandes', *XVIII Congr. Intern. Géogr., Rio de Janeiro* **ii**, 155–62.

GELLERT, J. (1955) 'Morphologische Probleme im Rumpftreppengebirge und Schichtstufenland', *Wiss. Zeitschr. Pädagog. Hochschule Potsdam, Math.-Naturw. Reihe*, **2**, 65–80.

GILEWSKA, S. (1963) 'Relief of the mid-triassic escarpment in the vicinity of Bedzin', *Prace Geogr.* no. 44.

HEMPEL, L. (1957) 'Flächenformen und Flächenbildung in der Stufenland-schaft. Beobachtungen aus Deutschland und Grossbritanien', *Peter-manns Mitt.* **101**, 178–84.

HEMPEL, L. (1958) 'Probleme der Oberflächenformung in Grossbritanien unter klimamorphologischer Fragestellung', *Petermanns Mitt.* **102**, 13–27, 81–9.

JUNGERIUS, P. (1964) 'The upper coal measures cuesta in eastern Nigeria', *Z. Geomorph.* suppl. 5, pp. 167–76.

MENSCHING, H. (1964) 'Die regionale und klimatisch-morphologische Differenzierung von Bergfussflächen auf der Iberischen Halbinseln', *Würzburger Geogr. Arb.*, no. 12, pp. 141–58.

MONOD, T. (1952) *L'Adrar mauritanien*. 2 vols., Bull. Dir. Mines A.O.F.

MORTENSEN, H. (1949) 'Rumpffläche, Stufenlandschaft, alternierende Abtragung', *Petermanns Mitt.* pp. 1–14.

MORTENSEN, H. (1953) 'Neues zum Problem des Schichtstufenlandschaft. Einige Ergebnisse einer Reise durch den Südwestern der USA, Sommer und Herbst 1952', *Dtscher. Geogr. Tag*, Essen, pp. 113–14.

SCHMIDT, E. (1954) 'Geomorphologische Studien im hinteren Odenwald und im Bauland', *Forsch. zur Dtschen Landeskunde*, **86**.

SCHMITTHENNER, H. (1956) 'Probleme der Schichtstufenlandschaft', *Marburger Geogr. Arb.* no. 3, 87 pp.

SUCHEL, A. (1954) 'Studien zur quartären Morphologie des Hilsgebietes', *Göttinger Geogr. Abh.* no. 17.

TRICART, J. (1954) 'La Champagne, La Lorraine (C.R. XXXVI Excur-sion géographique inter-universitaire)', *Ann. de Géogr.* **62**, 1–21, 88–98.

VERBEEK, T. (1962) 'De geomorfologie van de westerlijke Itüri', *Natuurwet. Tijdschr. (Gent)*, **54**, 177–200.

Shield margins

AWAD, H. (1949) 'La surface prénubienne dans le Sinaï montagneux central (partie Ouest)', *Rapport Comm. Surf. d'Aplanissement, Congr. Intern. Géogr.*, *Lisbonne*, pp. 149–54.

BEAUJEU-GARNIER, J. (1956) 'Carte morphologique du Bassin de Paris sous la direction d'A. Cholley, II: La région du Sud-Est', *Mém. et Doc. Centre Document. Cartogr. Centre National des Recherches Scientifiques*, **5**, 59–66.

BOMER, B. (1956b) 'Carte morphologique du Bassin de Paris, sous la direction d'A. Cholley, III: Le Sud et le Sud-Est du Bassin Parisien', *Mém. et Doc. Centre Document. Cartogr. CNRS*, **5**, 67–78.

DAUTRY, M. (1953) 'Sur la morphogénèse des reliefs structuraux de la bordure septentrionale du Massif Central Français,, *Bull. Soc. géol. Fr.*, pp. 146–50.

DEMANGEOT, J. (1961) 'Pseudo-cuestas de la zone intertropicale', *Bull. Ass. Géogr. fr.* **296/7**, 2–16.

DRESCH, J. (1959) 'Notes sur la morphologie de l'Aïr', *Bull. Ass. Géogr. fr.* pp. 2–20.

Bibliography

JOURNOT, C. (1948) 'Le couloir périphérique des Maures, étude morphologique', *Ann. de Géogr.* **57**, 97–108.
MOREAU, G. (1953) 'Le Boischaut. Etude morphologique régionale', *Bull. Ass. Géogr. fr.*, pp. 19–30.
STEFFEN, M. (1951) 'Zur Morphologie des südlichen Randgebietes der Luxemburger-Ardennen', *Serv. Géol. Luxembourg*, no. 8.
VOGT, J. (1962b) 'Notes de géomorphologie gabonaise', *Rev. Géomorph. dyn.* **13**, 161–9.

Chapter 4   Faults and volcanoes

Only works complementary to geological treatments of faults and volcanoes are listed here.

Faults

Tectonic aspects

ALLEN, C. R. (1962) 'Circum-Pacific faulting in the Philippines-Taïwan regions', *Journ. Geophys. Res.* **67**, 4795–812.
CAILLEUX, A. (1958) 'Etude quantitative des failles', *Rev. Géomorph. dyn.* **9**, 129–45.
CAILLEUX, A. (1960) 'Etude quantitative de petites failles au Nord de la Rochelle (Charente-Maritime)', *Rev. Géomorph. dyn.* **11**, 106–12.
CLARK, —. 'Classification of faults', *Bull. Am. Ass. Petrol. Geol.* **27**, 1245–65, 1633–42.
DRAKE, C. L. and GIRDLER, R. W. (1964) 'A geophysical study of the Red Sea', *Geophys. J. Royal Astron. Soc.* **8**, 473–95.
DUFLOS, H. (1960) 'Etude statistique des failles du bassin houiller de Liège', *Cahiers Géol.* **58/61**, 582–8.
ILLIES, H. (1962) 'Prinzipien und Entwicklung des Rheingrabens, dargestellt am Grabenabschnitt von Karlsruhe', *Mitt. Geol. Staatinst. Hamburg*, **31**, 58–122.
SCHENK, E. (1955) 'Postpliozäne Krustenbewegungen mit Faltenformen in der Wetterau', *Geol. Rdsch* **43**, 93–103.
WEBER, C. (1961) 'Structure géologique de la Californie: revue des problèmes actuels', *Rev. Géogr. phys. et Géol. dyn.* new ser. **4**, 67–88.

Faults in action

COTTON, C. A. (1951a) 'Une côte de déformation transverse à Wellington (Nouvelle-Zélande)', *Rev. Géomorphol. dyn.* **2**, 97–109.
GRANDAZZI, M. (1954) 'Le tremblement de terre des Iles Ioniennes (août 1953)', *Ann. de Géogr.* **63**, 431–53.
HOMORODI, L. (1962) 'Untersuchungen der rezenten Erdkrustenbewegungen in Ungarn. Symp. Erdkrustenbewegungen Leipzig', *Abh. Dtschen. Ak. Wiss. Berlin*, pp. 92–100.

KOBAYASHI, K. (1954) 'A short report on the geomorphic history and the pleistocene geology of the Matsumoto basin and its adjoining mountains', *Journ. Shinshu. Univ.* **4**, 87–98.

KOBAYASHI, K. (1955) 'Up and down movements now in action in Japan', *Geol. Rdsch.* **43**, 233–47.

NESTEROFF, W. (1959) 'Age des derniers mouvements du graben de la Mer Rouge déterminés par la méthode du C14 appliquée aux récifs fossiles', *Bull. Soc. Géol. Fr.*, 7th ser. **1**, 415–18.

RANTSMAN, E. Y. 'Neo and present tectonics of seismic areas in mountainous Middle Asia according to geomorphological observations', *Ac. Sc. URSS, Moscou, Commission de Géogr.* (Recent tectonic movements . . .), pp. 135–49.

SAINT-AMAND, P. (1961) 'Observaciones e interpretaciôn de los terremotos chilenos de 1960', *Comunic. Esc. Geologia, Univ. de Chile*, **i**, no. 2.

SOLONENKO, V.-P. (1950) 'Phénomènes dynamiques en rapport avec la néotectonique de la Sibérie orientale', *C.R. Ac. Sc. URSS, Sér. Géogr.* **72**, 109–12.

STEVENS, G. (1956) 'Earth movements in the Wellington area', *N.Z. Geogr.* **12**, 189–94.

WEBER, C. (1962) 'Les failles à déplacement horizontal dans la tectonique japonaise', *Rev. Géogr. phys. et Géol. dyn.* new ser. **5**, 131–41.

WEGMANN, E. (1955) 'Tectonique vivante: vue d'ensemble sur les travaux de la réunion de printemps 1954 à Mayence', *Geol. Rdsch.* **43**, 273–306.

Studies in regional geomorphology

CAHEN, L. (1954) *Géologie du Congo Belge*. Liège, Vaillant-Carmanne.

COTTON, C. A. (1949) 'Tectonic scarps and fault valleys', *C.R. Congr. Intern. Géogr., Lisbonne*, **ii**, 191–200.

COTTON, C. A. (1950a) 'Quelques aspects du relief de failles', *Rev. Géomorph. dyn*, **1**.

COTTON, C. A. (1950b) 'Tectonic relief', *Ann. Ass. Am. Geogr.* **40**, 181–7.

COTTON, C. A. (1950c) 'Tectonic scarps and fault valleys', *Bull. Geol. Soc. Amer.* **61**, 717–58.

COTTON, C. A. (1957a) 'An example of transcurrent-drift tectonics', *New Zealand Journ. Sc. and Techn.*, B, **38**, 939–42.

COTTON, C. A. (1951b) 'Fault valleys and shutter ridges at Wellington', *N.Z. Geogr.* **7**, 62–8.

COTTON, C. A. (1953) 'Tectonic relief, with illustrations from New Zealand', *Geogr. J.* **119**, 213–22.

COTTON, C. A. (1957b) 'Tectonic features in a coastal setting at Wellington', *Trans. Roy. Soc. N.Z.*, pp. 761–90.

DIXEY, F. (1956) 'The East African rift system', *Col. Geol. and Mining Res., Bull., Suppl.* no. 1, London.

FEIO, M. and SOARES DE BRITO, R. (1949) 'Les vallées de fracture dans le modelé granitique portugais', *C.R. Congr. Intern. Géogr., Lisbonne*, **ii**, 254–62.

GELLERT, J. (1954) 'Das grosse ostafrikanische Grabensystem', *Z. Erdk-Unterr.* **6**, 225–36.

HAMMOND, E. (1954) 'A geomorphic study of the Cape Region, Baja California', *Univ. of California, Publ. in Geogr.* **10**, no. 2, 45–112.

HEINZELIN, J. DE (1955) *Le Fossé tectonique sous le parallèle d'Ishango*. Brussels Inst. Parcs Nat. Congo Belge, Bruxelles, 148 pp.

HEINZELIN, J. DE (1962) 'Les formations du Western Rift et de la cuvette congolaise', *Actes IVe Congr. Panafr. Préhist. et Quat., Musée Royal Tervuren, Ann., Sér. in-8. Sc. Hum.* no. 40, 219–43.

HUETZ DE LEMPS, A. (1955) 'Le relief de la Nouvelle-Zélande', *Rev. Géogr. alp.* **43**, 5–96.

JOURNAUX, A. (1956) 'Les plaines de la Saône et leurs bordures montagneuses', Caen, 532 pp.

LEPERSONNE, J. (1956b) 'Les aplanissements d'érosion du Nord-Est du Congo belge et des régions voisines', *Ac. Sc. Col., Bruxelles, Sc. Nat., Mém., NSN*, no. 7, 109 pp.

POUQUET, J. (1964) 'La capture du Furnace Creek Wash par le Gower Gulch (Vallée de la Mort, Californie)', *Bull. Ass. Géogr. fr.* **322–3**, 43–51.

RUHE, R. (1956) 'Landscape evolution in the high Ituri, Belgian Congo', *Publ. INEAC*, Sér. Sc., no. 66.

TEALE, E. O. (1949) 'The river system in Tanganyika in relation to tectonic movement', *C.R. Congr. Intern. Géogr., Lisbonne*, **ii**, 233–42.

VAUMAS, E. DE (1961) 'Structure et morphologie du Proche-Orient. Nouvel essai de synthèse et orientations de recherche', *Rev. Géogr. alp.* **49**, 433–509.

VERSTAPPEN, H. T. (1959) 'Geomorphology and crustal movements of the Aru Island in relation to the pleistocene drainage of the Sahul Shelf', *Am. J. Sci.* **257**, 491–502.

VONFELT, J. (1955) 'La bordure vosgienne entre Sélestat et Rouffach', *Rev. Géomorph. dyn.* **6**, 7–33.

Volcanism

There is an enormous literature on this subject, of interest to geologists, geomorphologists and geophysicists. The items listed here give an idea of the range of problems covered.

General works

COTTON, C. A. (1944) *Volcanoes, as landscape forms*, Christchurch (N.Z.), Whitcombe and Tombs.

EATON, J. P. and MURATA, K. S. (1960) 'How volcanoes grow', *Science,* **132**, 925–38.

GREEN, J. and SHORT, N. M., eds. *Volcanic Landforms and Surface Features*, Berlin, Springer-Verlag.

OLLIER, C. D. (1969) *Volcanoes*, New York, M.I.T. Press.

RITTMANN, A. (1962) *Volcanoes and their Activity*, Wiley.

Regional monographs featuring several types of volcanoes

BIAYS, P. (1956) 'Introduction à la géomorphologie du Sud-Ouest de l'Islande', *Ann. Litt. Univ. Besançon*, **13**, 118 pp.

BOUT, P. (1953) *Etudes de géomorphologie dynamique en Islande. Expédition Polaire français*. Paris, Hermann.

BOUT, P. (1960) *Le Villafranchien du Velay et du bassin hydrographique moyen et supérieur de l'Allier. Corrélations françaises et européenes*. Le Puy.

BOUT, P., DERRUAU, M., DRESCH, J. and PEGUY, CH. P. (1961) 'Observations de géographie physique en Iran septentrional', *Mém. Centre Document. Cartogr. CNRS*, **8**, 9–101.

DERRUAU, M., ESTIENNE, P. and FEL, A. (1958) 'Etat de nos connaissances géographiques sur le Massif Central Français', *Rev. d'Auvergne*, **72**, 64 pp.

HUETZ DE LEMPS, A. (1955) 'Le relief de la Nouvelle-Zélande', *Rev. Géogr. alp.* **43**, 5–96.

KREJCI-GRAF, K. (1956) 'Vulkanologische Beobachtungen auf den Azoren', *Frankfurter Geogr.* **30**.

LASSERRE, G. (1961) *La Guadeloupe, la nature et les hommes*, vol. i, Bordeaux.

RUDEL, A. (1962) *Les Volcans d'Auvergne*, Clermont-Ferrand.

THORARINSSON, S., EINARSSON, T. and KJARTANSSON, G. (1959) 'On the geology and geomorphology of Iceland', *Geogr. Annlr.* **41**, 135–69.

TRICART, J. (1961) 'Aperçu sur le Quaternaire du Salvador (Amérique Centrale), *Bull. Soc. géol. Fr.* 7th ser. **3**, 59–68.

VERSTAPPEN, H. T. (1963) 'Geomorphological observations on Indonesian volcanoes', *Tijdschr. K. nedl. aardrijksk. Genoot.* **80**, 237–51.

VLODAVEC, V. J. (1954) *Die Vulkane der Sowjetunion.*, trans. H. Täubert. Gotha, Geogr. Kartogr. Anstal.

VOSKRESSENSKY, S. S. and DOUMITRACHKO, N. (1956) 'Caractéristiques comparatives des reliefs des régions volcaniques (basaltiques) de l'URSS', *Essais de Géogr. Ac. Sc. URSS*, Moscow, pp. 93–103.

WILLIAMS, H. (1950) 'Volcanoes of the Paricutin region (Mexico)', *U.S. Geol. Surv., Bull.* **965** B, Washington, pp. 165–271.

Descriptions of volcanic phenomena

BELLAIR, P. (1960) 'L'éruption du Kilauea (Hawaï) de novembre 1959 et janvier-février 1960', *C.R. Somm. Soc. Géol. Fr.*, pp. 79–80.

BORDET, P. and TAZIEFF, H. (1963) 'Remarques sur l'éruption du Katmaï et de la Vallée des Dix Mille Fumées et sur le problème des ignimbrites', *Bull. Soc. géol. Fr.*, 7th ser. **5**, 210–13.

Bibliography

COATS, R. (1950) 'Volcanic activity in the Aleutian Arc', *U.S. Geol. Surv.,
Bull.* 974 B, 35–49.

GEZE, B. (1963b) 'Quelques enseignements des éruptions d'Hawaï en
1959–60', *Bull. Soc. géol. Fr.* 7th ser. **5**, 176–81.

REMY, J.-M. (1963) 'L'éruption volcanique de 1960 au Lopévi (Nouvelles
Hébrides)', *Bull. Soc. géol. Fr.* 7th ser. **5**, 188–97.

RIBEIRO, O. (1954) 'A Ilha do Fogo e as suas erupçoês', *Mem. Junta
Investig. Ultramar.*

SEGERSTROM, K. (1960) 'Erosion and related phenomena at Paricutin in
1957', *U.S. Geol. Surv., Bull.* 1104 A, 18 pp.

THORARINSSON, S. (1956) *Hekla on fire.* Munich, Hans Reich.

WILCOX, R.-E. (1959) 'Some effects of recent volcanic ash falls with special
reference to Alaska', *US Geol. Surv., Prof. Paper* 1028 N, pp. 409–76.

Monographs

Explosive types, ignimbrites, nuées ardentes

DERRUAU, M. (1962) 'Etudes sur quelques volcans explosifs japonais',
Bull. Ass. Géogr. fr. no. 307/8, pp. 156–68.

DORN, P. (1948) 'Ein Jahrhundert Riesgeologie', *Z. dtn. Geol. Ges.* C,
pp. 348–65.

GENZ, H. (1955) 'Die Maare der Eifel', *Zeitschr. für den Erdkundeunterr.* **7**,
79–85.

GEZE, B., HUDELEY, H. VINCENT, P. and WACRENIER, P. (1959) 'Les
volcans du Tibesti. (Sahara du Tchad)', *Bull. Volcanol.* 2nd ser. **22**,
135–72.

KUNTZ, P. (1955) 'Sur les volcans d'explosion au Maroc oriental', *C.R.
Somm. S.G. Fr.* pp. 338–340.

MCCALL, G. J. H. (1963) 'Classification of calderas "Krakatoan" and
"Glencoe" types', *Nature,* **197**, 136–8.

RITTMANN, A. (1962b) 'Erklärungsversuch zum Mechanismus der
Ignimbritausbrüche', *Geol. Rdsch,* **52**, 853–61.

ROSS, C. S. and SMITH, R. L. (1961) 'Ash-flow tuffs: their origin, geologic
relations, and identification', *U.S. Geol. Surv., Prof. Paper* 366, pp. 1–80.

Laccoliths

MILLOT, G. and DARS, R. (1959) 'L'archipel des îles de Los: une structure
annulaire subvolcanique (Rép. de Guinée)', *Notes du SGPM, Dakar,*
no. 2, pp. 49–56.

MONOD, T. (1951) 'Accidents circulaires et cratériformes ouest-sahariens',
C.R. Somm. S. G. Fr. pp. 120–2.

PASOTTI, P. (1957) 'Los domos lacoliticos de Tandil (Prov. de Buenos-
Aires)', *Univ. Nac. Lit. Rosario, Fac. de Ciencias, Public.* **42**, 71 pp.

Various

BOZIYAN, X. A. (1956) 'Les éruptions des volcans de boue en Azerbaïdjan', *Priroda*, **8**, 91–3 (in Russian).

BOZON, P. (1963) 'Contribution à l'étude des formes volcaniques de l'Ardèche', *Rev. Geogr. alp.* **51**, 591–674.

DERRUAU, M. and PEGUY, CH. P. (1953) 'Aspects et problèmes de l'Etna', *Bull. Ass. Géogr. fr.* pp. 88–101.

GUILCHER, A. (1950) 'Définition d'un type de volcan "écossais"', *Bull. Ass. Géogr. fr.* pp. 1–10.

MENSCHING, H. (1957) 'Geomorphologie der Hohen Rhön und ihres südlichen Vorlandes', *Würzburger Geogr. Arb.* no. 4/5, pp. 47–88.

OLLIER, C. D. (1964) 'Caves and related features of Mount Eccles', *Victoria Nat. (Melbourne)*, **81**, 64–71.

SATO, H. (1950) 'Geomorphological classification of lava flows', *Bull. Geogr. Inst., Tokyo*, no. 1.

SNEAD, R. E. (1964) 'Active mud volcanoes of Baluchistan, West Pakistan', *Geogr. Rev.* **54**, 546–60.

WILLIAMS, L. A. J. (1963) 'Lava tunnels on Suswa Mountain, Kenia', *Nature*, **199**, 348–50.

MULLINEAUX, D. R. and CRANDELL, D. (1962) 'Recent "lahars" from Mount Helens, Washington', *Bull. Geol. Soc. America*, **73**, 855–70.

Index

The more important references are in **bold** type

303